A User's Guide to the Gottman-Williams
Time-Series Analysis Computer Programs for Social Scientists

A User's Guide
to the Gottman-Williams
Time-Series Analysis Computer Programs
for Social Scientists

ESTHER A. WILLIAMS
JOHN M. GOTTMAN

University of Illinois, Champaign

CAMBRIDGE UNIVERSITY PRESS

Cambridge
London   New York   New Rochelle
Melbourne   Sydney

CAMBRIDGE UNIVERSITY PRESS
Cambridge, New York, Melbourne, Madrid, Cape Town, Singapore,
São Paulo, Delhi, Dubai, Tokyo, Mexico City

Cambridge University Press
The Edinburgh Building, Cambridge CB2 8RU, UK

Published in the United States of America by
Cambridge University Press, New York

www.cambridge.org
Information on this title: www.cambridge.org/9780521280594

First published 1982
Reprinted, 1983

A catalogue record for this publication is available from the British Library

Library of Congress Cataloguing in Publication Data

Williams, Esther A.
A User's Guide to the Gottman-Williams
Time-Series Analysis Computer Programs for Social Scientists.
I. Time-series analysis – Computer programs.
I. Gottman, John Mordechai. II. Title.
HA30.3.W54   519.5 5′028542   81-10058
ISBN 0 521 28059 I   AACR2

ISBN 978-0-521-28059-4 Paperback

Contents

Introduction   *page*   1

Package 1. Univariate model fitting   7
   DETRND   9
   LINFIL   17
   DESINE   23
   ARFIT   29
   SPEC   39

Package 2. Applications of univariate model fitting   49
   FORCST   51
   ITSE   57

Package 3. Multivariate time-series analysis   65
   BIVAR   67
   TSREG   85
   CRSPEC   93

Contents

Introduction

These programs and this manual were written with beginners in mind. We have a lot of sympathy for researchers who have an idea of what they want to learn from their data and who hope for easily intelligible output. They would like a set of flexible, powerful programs that they can grow into as they learn about time-series analysis. They would be likely to avoid using programs that require a great deal of technical knowledge even to punch the control cards. For example, consider researchers who think their data may be cyclic. They skim a few chapters in a time-series book and learn that there is a thing called a spectral density function, and if that has a statistically significant peak, their data has a cycle at that particular frequency. Then they read the computer manual and find that they need to answer the question, "Which spectral window do you want to use? Punch 1 for Tukey-Hanning, 2 for Bartlett, 3 for Parzen, etc." They are likely to be discouraged from ever using the program if they have to answer perplexing questions such as this.

What we have done is to make these decisions for the user by limiting the number of options. We have done this only when it makes very little practical difference which option is chosen. We specify the equations we use so that an experienced user can modify the programs to suit special needs.

We think that these are straightforward programs to use. We also think that the output is useful for making statistical decisions.

Overview of time-series analysis

For a more detailed, but still nonmathematical, overview of time-series analysis see Chapters 1 to 7 of Gottman's book *Time-Series Analysis*. There are two kinds of time-series analyses, *time-domain* and *frequency domain*. Time-domain analysis is useful for such tasks as forecasting and frequency domain analysis is useful for such tasks as detecting cyclicity. However, they really ought be go hand in hand. For example, in building a good time-domain model it is useful to know if there are cyclic, or "seasonal" components and this is accomplished in part by a frequency-domain analysis.

One of the things researchers may wish to do is to fit a model to one time-series. The model breaks the data into understandable components such as linear trend, cycles, and a stochastic stationary time-series. There are a variety of applications of model fitting. These include forecasting and testing the effectiveness of interventions. Another kind of time-series analysis is multivariate. This involves assessing the relationships among two or more time-series. Once again,

1

this can be done in either the time domain or the frequency domain.

Overview of the programs

There are three packages of programs described in this manual. The first package is for *univariate model fitting.* It is described in the box below.

PACKAGE 1: UNIVARIATE MODEL FITTING

A. TIME DOMAIN

1. *Removing Trend.* Program DETRND removes linear trend. It tests the significance of the trend line parameters and outputs the residual time-series, which is the deviations of the original series around the trend line.

2. *Linear Filter.* Program LINFIL is useful for transforming a time-series in a variety of ways, including differencing, smoothing, and removal of a curve (see the Porges example in Chapter 9 of Gottman's book).

3. *Removing Cycles.* Program DESINE will remove a cycle of any period specified by the user. It performs a least-squares fit that finds the best amplitude and phase parameters. The output is a test of the significance of the cycle removed and a residual time-series.

4. *Autoregressive Fit.* Program ARFIT performs an autoregressive model fit up to the order specified by the user. It outputs the variance of the series, its autocovariances, autocorrelations, partial autocorrelation coefficients, the autoregressive model coefficients, tests of their significance, the residual from the model fit and a Box-Pierce test of whether the residual series is white noise.

B. FREQUENCY DOMAIN

Spectral Density. Program SPEC computes the autocovariances, autocorrelations, spectral density estimates (using a Tukey-Hanning window), confidence intervals for each frequency, and one test of significance that compares the spectral density to the theoretical density of white noise. The user must specify only the maximum lag for the calculations. Larger lags divide the frequency range into a finer grid. This program also can compute the periodogram. The density functions and confidence intervals are plotted as well as printed.

The second package is for two applications of univariate model fitting, *forecasting* and *assessing the effects of an intervention*. The second package is described in the box below.

PACKAGE 2: APPLICATIONS OF UNIVARIATE MODEL FITTING

1. *Forecasting.* Program FORCST computes the best one-step ahead prediction using the autoregressive model fitting process, and the one-step ahead prediction variance.

2. *Intervention Effects.* Program ITSE, which stands for Interrupted Time Series Analysis performs an analysis for change in level and slope parameters following an intervention. It uses the linear autoregressive model procedure described in Chapter 26 of Gottman's book. The user must supply only the order of the autoregressive model. The output includes pre- and post-intervention slope, intercept parameters, model parameters, significance tests of the parameters, an overall F-test, and t-tests for change in slope and intercept.

The third package is for *multivariate time-series analysis*. This package is described in the box below.

PACKAGE 3: MULTIVARIATE TIME-SERIES ANALYSIS

A. TIME DOMAIN

1. *Bivariate Time Domain.* Program BIVAR performs the Gottman-Ringland procedure described in Chapter 25 of Gottman's book. This analysis examines lead-lag relationships between two time-series, controlling for auto-correlation in each series. When it is used twice, once with each series as input, it can assess asymmetrical or bidirectional influence between the series. The user must specify the maximum lag values for auto and cross relationship between the two time-series. Chapter 25 describes how to obtain guesses for these values using spectral and cross-spectral analyses. The output is extensive and will not be described here. A summary table provides a chi-square test of the adequacy of the step-wise procedure the program uses as well as a test of association between the two series, with one series as

input to the second.

2. *Time-Series Regression.* Program TSREG performs a two-stage generalized least-squares regression. This predicts one criterion time-series Y_t from a set of other predictor time-series X_{1t}, X_{2t}, $\cdots$, X_{kt}. The predictor series can be lagged, prewhitened, and so on using program LINFIL.

B. FREQUENCY DOMAIN

Cross-Spectral Analysis. Program CRSPEC performs a bivariate frequency domain time-series analysis. The user supplies only the maximum lag for the analysis. The program outputs the spectral density function for each series, the autocovariances, the cross-covariances, the coherence spectrum, with a significance test for zero coherence, and the phase spectrum, with confidence intervals. The spectra are all also plotted. The user can specify a shift parameter for the analysis (see Chapter 23, Gottman's book).

What the programs do

The chart below is an overview of the programs from a slightly different point of view.

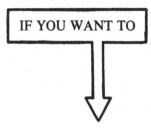

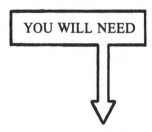

1. Assess the effects of an intervention, i.e., do an interrupted time-series analysis.

1. Program ITSE. It would be useful to employ the univariate model-fitting programs for examining and removing trend and seasonal components [programs DETRND, LINFIL, DESINE, ARFIT, and SPEC].

2. Examine one time-series for component cyclicities.

2. Program SPEC (also can be used to compute the periodogram).

3. Forecast data (limited to one step ahead).

3. Program FORCST.

4. Examine two series for cyclical covariation, i.e., synchronicity and cyclic lead-lag relationships.

4. Program CRSPEC.

5. Perform bivariate time-series analysis, controlling for autocorrelation in each series (Gottman-Ringland procedure).

5. Program BIVAR.

6. Do regression from a set of predictor time-series to a criterion time-series.

6. Program TSREG.

Package 1.
Univariate model fitting

DETRND

Program description

The program DETRND is a least-squares linear regression program. In this program, the least-squares straight line is determined for a given set of observations; hypothesis tests are performed on the slope and intercept of the regression line; and the vector of errors of the estimate is computed.

Least-squares linear regression is described in more detail in Chapter 14 of Gottman's *Time-Series Analysis*.

Program steps

In the program, the mean of the N observations (Y_i) is calculated as:

$$\bar{Y} = \frac{1}{N}\sum_{i=1}^{N} Y_i = \hat{a} \ .$$

The variance is determined by:

$$s^2 = \frac{1}{N-2}\sum_{i=1}^{N} (Y_i - \bar{Y})^2 \ .$$

The value $\sum x_i^2$, which is the sum of the squared deviations from the mean of the fixed points in time of the observations is calculated as:

$$\sum x_i^2 = \frac{N(N+1)(N-1)}{12} \ .$$

This value is used in the calculation of the slope of the line:

$$\hat{b} = \frac{\sum_{i=1}^{N} iY_i - \frac{N(N+1)}{2}\bar{Y}}{\sum x_i^2} \ .$$

The equation of the least-squares regression line is:

$$\hat{Y}_i = \hat{a} + \hat{b}\left(i - \frac{N+1}{2}\right) \ .$$

Student's t for the slope is determined for hypothesis testing as:

$$t(\hat{b}) = \frac{\hat{b}}{\sqrt{s^2/\Sigma x_i^2}} \; .$$

Degrees of freedom for testing are $N-2$. Similarly, Student's t for $\hat{a}$ is:

$$t(\hat{a}) = \frac{\hat{a}}{\sqrt{s^2/N}} \; .$$

Again, the degrees of freedom for testing are $N-2$.

When all of the estimates $(\hat{Y}_i)$ have been calculated with the equation of the least-squares line, the vector of the residuals (errors of estimate) (E_i) is calculated by:

$$e_i = y_i - \hat{y}_i \; .$$

Program input

1. First input card.
   - Cols. 2-5   Number of data points (N). Maximum is 1500.
   - Col. 20   Number of format cards to follow. Maximum is 5.
   - Col. 25   Set to 1 if vector of residuals (errors of estimates) is to be punched.
2. Second input card.
   - Cols. 1-80   The title to be printed for identification of output.
3. Third input card or cards.
   - Cols. 1-80   Data format is specified on this card or cards. Standard FORTRAN F-format is required, enclosed in parentheses.
4. Data cards.
   - Cols. 1-80   Data cards follow the last format card. The data are punched in the format specified on the format card or cards.

Program output

When the program is run, the following information is output:
1. Program name: DETRND PROGRAM.
2. The title for the printout identification.
3. Number of data points specified (N).

4. Number of format cards specified.
5. Data format is repeated back to user.
6. The first and last data points are printed for user to check format.
7. Intercept of fitted line ($\hat{a}$).
8. Slope of fitted line ($\hat{b}$).
9. Equation of fitted least-squares regression line.
10. Student's t for slope with degrees of freedom.
11. Student's t for intercept with degrees of freedom.
12. The error of estimate vector (E_i) with an error value for each data point i. The vector of errors is punched one error to a card if requested on first input card.
13. End-of-program message.

Special notes

1. If N is desired to be greater than 1500, the dimensions of the Y array and the ERROR array will have to be modified in the program as well as the statements near the beginning of the program which check for the maximum value of N.
2. In the program, output for the punch is written to unit number 7. Other input and output will be on the default system devices specified for input and output (usually the card reader and printer). A CDC computer will write unit 7 to a file called PUNCH which may have to be punched as a separate step. An IBM computer will punch directly if the card punch device is defined as unit 7.
3. None of the computation in the program is machine specific, so going from one computer to another should not create any special problems. The language used is FORTRAN IV. The user may need to remove the first card of the program (the PROGRAM card) and change double quotation marks to single quotation marks in all FORMAT statements.
4. If the user wishes to have input read from other than the default system device, the device can be redefined for the execution step and data specification and format cards must precede the data on that device. This works nicely at a timesharing computer terminal. Or if the user wishes to read only data from another device, the program statement or statements which read data will have to be modified. In all programs, all data is read in only once just after the comment card "READ IN DATA."

Example

Sample input

```
    60                1     1
SAMPLE RUN OF LINEAR REGRESSION PROGRAM
(F4.2)
1.61
1.00
1.38
2.05
2.21
2.00
1.35
2.17
2.25
2.71
3.39
2.50
3.56
4.00
3.76
3.29
3.28
2.31
2.50
2.33
2.27
2.35
3.33
1.33
2.19
1.44
1.24
1.44
4.40
4.50
3.56
3.00
2.29
1.83
2.24
1.39
2.82
3.29
3.33
3.33
2.71
2.29
1.38
2.83
1.00
1.33
3.28
3.12
3.59
3.24
```

```
2.56
1.44
1.17
2.29
4.00
4.47
4.60
3.37
2.42
2.20
```

Sample output

```
                 DETRND PROGRAM

SAMPLE RUN OF LINEAR REGRESSION PROGRAM

   60 DATA POINTS
    1 FORMAT CARDS
   DATA FORMAT:  (F4.2)
   FIRST DATA POINT IS    1.6100
   LAST DATA POINT IS     2.2000
   RESIDUAL (ERROR) SERIES WILL GO TO PUNCH.   PUNCH FORMAT:   (I7,F10.3)

INTERCEPT OF FITTED LINE = A =     2.575
SLOPE OF FITTED LINE     = B =      .011

EQUATION OF FITTED LINE IS:
             Y(I) =   A   +(    B   (I-(N+1)/2))
OR:          Y(I) = 2.575 +(  .011(I -  30.5))

STUDENT'S T FOR SLOPE:
             T(B) =    1.583        DF =  58

STUDENT'S T FOR INTERCEPT:
             T(A) =   20.642        DF =  58

        ERROR SERIES (DIFFERENCE BETWEEN DATA AND FITTED LINE)
  I     ERROR   I    ERROR   I    ERROR   I    ERROR   I    ERROR
  1    -.629    2  -1.250    3   -.882    4   -.223    5   -.074
  6    -.296    7   -.957    8   -.149    9   -.080   10    .369
 11    1.037   12    .136   13   1.184   14   1.613   15   1.362
 16     .880   17    .859   18   -.123   19    .056   20   -.125
 21    -.197   22   -.128   23    .840   24  -1.171   25   -.322
```

26	-1.084	27	-1.295	28	-1.107	29	1.842	30	1.931
31	.979	32	.408	33	-.314	34	-.785	35	-.386
36	-1.248	37	.171	38	.629	39	.658	40	.647
41	.015	42	-.416	43	-1.338	44	.101	45	-1.740
46	-1.422	47	.517	48	.345	49	.804	50	.442
51	-.249	52	-1.380	53	-1.662	54	-.553	55	1.145
56	1.604	57	1.723	58	.481	59	-.480	60	-.712

END OF DETRND PROGRAM

LINFIL

Program description

The program LINFIL is the linear filter program. The program performs the transformation:

$$Y_t = \sum_{i=0}^{k} a_i X_{t-i}$$

on time-series X to form the new filtered series Y as described in Chapter 3 of Gottman's *Time-Series Analysis*.

To *smooth* a series using a moving average, the a_i are the weights. For example, a simple three-point moving average is $a_i = \frac{1}{3}$, $k = 3$. To difference a series once, the weights are $a_0 = 1$, $a_1 = -1$. To difference a series twice, $a_0 = 1$, $a_1 = -2$, $a_2 = 1$. To difference a series three times, $a_0 = 1$, $a_1 = -3$, $a_2 = +3$, and $a_3 = -1$. These numbers were obtained by expanding $(1-B)^d$, where d is the order of differencing (see Chapter 9 of Gottman's book).

Program steps

The program user supplies the time-series (X) and a set of weights or coefficients (a) for the filter. N is the number of data points in the time-series, and NC is the number of coefficients or weights to be applied.

The new filtered series (Y) is computed as:

$$Y_i = \sum_{j=1}^{NC} a_j X_m \,.$$

in which $i = 1, 2, \cdots, N-NC+1$ and $m = i+NC-j$. Notice that the first weight or coefficient applies to the most recent data point in time, the second weight to the next most recent data point, and so forth.

Program input

1. First input card.

Cols. 2-5	Number of data points (N) in the time-series. Maximum is 1500.
Cols. 7-10	Number of coefficients or weights (NC) in the filter.
Col. 20	Number of format cards for the time-series.

Maximum is 5.

Col. 22 Number of format cards for the filter coefficients. Maximum is 5.

Col. 25 Set to 1 if new series is to go to punch.

2. The second input card.

Cols. 1-80 The title to be printed on the output.

3. Next input card or cards.

Cols. 1-80 Time-series data format is specified on this card or cards. Standard FORTRAN F-format is used, enclosed in parentheses.

4. Time series data cards.

Cols. 1-80 The time-series data cards follow the last format card for the time-series. The data are punched in the format specified on the format card or cards.

5. Next input card or cards.

Cols. 1-80 The format for the coefficients or weights of the filter is specified on this card or cards. Again, standard FORTRAN F-format is used, enclosed in parentheses.

6. Coefficient or weight cards.

Cols. 1-80 The coefficients or weights of the filter follow the last format card, punched in the format specified by the coefficient format card or cards. For each calculation of a data point in the new series, the first weight entered is applied to the most recent data point in time in the original series, as described above.

Program output

When the program is run, the following information is output:

1. Program name: LINFIL PROGRAM.
2. The title for the printout identification.
3. Number of data points (N) specified.
4. Number of coefficients or weights in the filter (NC) specified.
5. Number of data format cards specified.
6. Number of coefficient format cards specified.
7. The data format is printed back for the user.
8. The first and last data points are printed for the user to check format.
9. The coefficient format is printed back for the user.
10. The first and last coefficient are printed for the user to check

format.
11. The new filtered series (Y) is printed.
12. The end of program message.

Special notes

1. If the number of data points in the time-series is required to be larger than 1500, the size of arrays X and Y will have to be changed and the statements which check for the maximum near the start of the program will have to be modified accordingly.
2. See Special Notes 2, 3 and 4 of program DETRND.

Example

Sample input

```
    60      5            1 1
SAMPLE RUN OF THE LINEAR FILTER PROGRAM
(F4.2)
1.61
1.00
1.38
2.05
2.21
2.00
1.35
2.17
2.25
2.71
3.39
2.50
3.56
4.00
3.76
3.29
3.28
2.31
2.50
2.33
2.27
2.35
3.33
1.33
2.19
1.44
1.24
1.44
4.40
4.50
3.56
3.00
2.29
1.83
2.24
```

20

```
1.39
2.82
3.29
3.33
3.33
2.71
2.29
1.38
2.83
1.00
1.33
3.28
3.12
3.59
3.24
2.56
1.44
1.17
2.29
4.00
4.47
4.60
3.37
2.42
2.20
(5F5.0)
  1.0  2.0  0.0 -2.0  1.0
```

Sample output

```
                 LINFIL PROGRAM

SAMPLE RUN OF THE LINEAR FILTER PROGRAM

      60 DATA POINTS
       5 COEFFICIENTS IN THE FILTER
       1 DATA FORMAT CARDS
       1 COEFFICIENT FORMAT CARDS
   DATA FORMAT:  (F4.2)
   FIRST DATA POINT IS      1.6100
   LAST DATA POINT IS       2.2000

   COEFFICIENT FORMAT:  (5F5.0)
   FIRST COEFFICIENT IS     1.0000
   LAST COEFFICIENT IS      1.0000
```

THE NEW SERIES

1	5.920	2	4.660	3	2.630	4	2.500	5	4.800
6	6.510	7	5.820	8	6.950	9	5.390	10	7.050
11	10.150	12	6.190	13	5.420	14	5.350	15	4.300
16	4.060	17	5.590	18	4.200	19	5.870	20	5.780
21	2.420	22	1.510	23	4.790	24	.870	25	6.590
26	12.260	27	10.920	28	2.760	29	3.690	30	3.790
31	3.460	32	4.290	33	4.230	34	6.280	35	9.370
36	5.740	37	5.610	38	4.340	39	2.630	40	3.500
41	4.790	42	2.860	43	1.660	44	10.510	45	8.170
46	5.190	47	6.080	48	2.500	49	1.160	50	2.750
51	8.260	52	11.570	53	10.130	54	6.860	55	4.220
56	2.310								

END OF LINFIL PROGRAM

DESINE

Program description

The program DESINE is the program to remove the seasonal components of a time-series. The procedure is described in Appendix 16.2 of Gottman's *Time-Series Analysis*.

Program steps

The user supplies the time-series **Y** with N points and the number of points, NP, in a season. The number of points in a season can be determined by spectral analysis.

The equation which expresses the components of each data point y_t is:

$$y_t = \hat{\mu} + \hat{A}\cos\theta\, t + \hat{B}\sin\theta\, t + e_t \, ,$$

in which $\theta = \dfrac{2\pi}{NP}$ and $t = 1, 2, \cdots, N$. The matrix **X** is constructed by the program as:

$$\begin{bmatrix} 1 & \cos\theta & \sin\theta \\ 1 & \cos2\theta & \sin2\theta \\ \vdots & \vdots & \vdots \\ 1 & \cos N\theta & \sin N\theta \end{bmatrix}$$

The least squares solution:

$$\hat{\beta} = (\mathbf{X}^T\mathbf{X})^{-1}\mathbf{X}^T\mathbf{Y}$$

yields the vector BETA with the 3 estimators $\hat{\mu}$, $\hat{A}$ and $\hat{B}$. In completing this computation, the Bauer-Reinsch method of matrix inversion is performed as described in Nash (1979).

The diagonal of the matrix $(\mathbf{X}^T\mathbf{X})^{-1}$ is the variance of each of the three estimators $\hat{\mu}$, $\hat{A}$ and $\hat{B}$.

The sum of squares for error (SSE) is calculated by the matrix equation:

$$SSE = \frac{(\mathbf{Y}-\mathbf{X}\hat{\beta})^T(\mathbf{Y}-\mathbf{X}\hat{\beta})}{N-3} \, .$$

The t-test is performed for each estimator:

23

$$t_\mu = \frac{\hat{\mu}}{SSE\ \sqrt{\text{1st diagonal in } (\mathbf{X}^T\mathbf{X})^{-1}}}$$

$$t_A = \frac{\hat{A}}{SSE\ \sqrt{\text{2nd diagonal in} (\mathbf{X}^T\mathbf{X})^{-1}}}$$

$$t_B = \frac{\hat{B}}{SSE\ \sqrt{\text{3rd diagonal in} (\mathbf{X}^T\mathbf{X})^{-1}}} .$$

In each case, degrees of freedom are $N-3$.

The residual series is:

$$\mathbf{E} = \mathbf{Y} - \mathbf{X}\hat{\boldsymbol{\beta}} .$$

Program input

1. First input card.

    Cols. 2-5   Number of data points (N) in the time-series. Maximum is 1000.

    Cols. 8-10   Number of points in a season (NP).

    Col. 20   Number of format cards to follow. Maximum is 5.

    Col. 25   Set to 1 if residual series is to go to a punch.

2. Second input card.

    Cols. 1-80   Title to be printed on output for later identification.

3. Next card or cards.

    Cols. 1-80   Time-series data format is specified on this card or cards. Standard FORTRAN F-format is used, enclosed in parentheses.

4. Time-series data cards.

    Cols. 1-80   The time-series data cards follow the last format card. The data are punched in the format specified on the format card or cards.

Program output

When the program is run, the following information is output:

1. Program name: DESINE PROGRAM.
2. Title of printout for identification.
3. Number of data points (N) specified.
4. Number of points in a season (NP) specified.
5. Number of data format cards specified.

6. The data format is printed back for the user.
7. If residual series is requested to be punched, the punch format will be printed.
8. The first and last data points are printed for the user to check format.
9. The estimators $\hat{\mu}$, $\hat{A}$ and $\hat{B}$ are printed with their t-test and degrees of freedom for the t-test.
10. The residual series is printed.
11. The residual series is punched if this was requested on the first input card.
12. End of program message.

Special notes

1. If the number of data points in the time-series is required to be larger than 1000, the size of arrays Y, X, TEMP, WORKA and WORKB in the program will have to changed and the statements which check for this maximum near the start of the program will have to be changed accordingly.
2. See Special Notes 2, 3 and 4 of the DETRND program.

Reference

Nash, J. C. (1979) *Compact numerical methods for computers: linear algebra and function minimisation.* New York: Halsted Press.

Example

Sample input

```
    60     9          1     1
SAMPLE RUN OF DESINE PROGRAM
(F4.2)
1.61
1.00
1.38
2.05
2.21
2.00
1.35
2.17
2.25
2.71
3.39
2.50
3.56
4.00
3.76
3.29
3.28
2.31
```

2.50
2.33
2.27
2.35
3.33
1.33
2.19
1.44
1.24
1.44
4.40
4.50
3.56
3.00
2.29
1.83
2.24
1.39
2.82
3.29
3.33
3.33
2.71
2.29
1.38
2.83
1.00
1.33
3.28
3.12
3.59
3.24
2.56
1.44
1.17
2.29
4.00
4.47
4.60
3.37
2.42
2.20

Sample output

DESINE PROGRAM

SAMPLE RUN OF DESINE PROGRAM

     60 DATA POINTS
     9 DATA POINTS IN A SEASON
     1 FORMAT CARDS
    DATA FORMAT: (F4.2)
    RESIDUAL SERIES WILL GO TO PUNCH
       PUNCHED FORMAT IS: (5(I5,F10.3))

THE FIRST DATA POINT IS   1.6100
THE LAST DATA POINT IS   2.2000

ESTIMATES

		T	DF
MU	2.546	27.961	57
A	-.523	-4.054	57
B	.455	3.543	57

THE RESIDUAL SERIES

N	E	N	E	N	E	N	E	N	E
1	-.828	2	-1.903	3	-1.821	4	-1.143	5	-.672
6	-.413	7	-.657	8	.317	9	.227	10	.272
11	.487	12	-.701	13	.367	14	1.118	15	1.347
16	1.283	17	1.427	18	.287	19	.062	20	-.573
21	-.931	22	-.843	23	.448	24	-1.083	25	.183
26	-.413	27	-.783	28	-.998	29	1.497	30	1.299
31	.367	32	.118	33	-.123	34	-.177	35	.387
36	-.633	37	.382	38	.387	39	.129	40	.137
41	-.172	42	-.123	43	-.627	44	.977	45	-1.023
46	-1.108	47	.377	48	-.081	49	.397	50	.358
51	.147	52	-.567	53	-.683	54	.267	55	1.562
56	1.567	57	1.399	58	.177	59	-.462	60	-.213

END OF DESINE PROGRAM

ARFIT

Program description

The program ARFIT is the autoregressive model fitting program. In this program, the initial variance, autocovariances and autocorrelations of original data are computed up to the order specified. Then the autoregressive coefficients, partial autocorrelation coefficients and the residual variance are computed for the autoregressive model fit of the desired order. The residual time series of the autoregressive fit is calculated along with the autocorrelations of the residual. The Box-Pierce statistic is calculated, as well as Student's t-ratio for significance testing of the autoregressive coefficients.

Autoregressive model fitting is described in more detail in Chapter 19 of Gottman's *Time-Series Analysis.*

Program steps

In the program, the mean of the original data is first calculated as:

$$\bar{X} = \frac{1}{T}\sum_{t=1}^{T} x_t ,$$

where T is the number of data points (x) in the original series. The variance is:

$$Var\ X = \frac{1}{T}\sum_{t=1}^{T}(x_t - \bar{X})^2 = C_0 .$$

The variance is autocovariance 0 (zero) and all other autocovariances (C_j) of the original data up to the maximum order specified are computed by:

$$C_j = \frac{1}{T}\sum_{t=1}^{T-j}(x_t - \bar{X})(x_{t+j} - \bar{X}) .$$

Usually the maximum order is chosen to be less than or equal to $(T/6)$ and for the autocovariances and autocorrelations, j goes from 1 to the maximum order chosen (M).

The autocorrelations (R_j) of the original data are calculated with the equation:

$$R_j = C_j/C_0 .$$

The coefficients (a) of the autoregressive model are computed next. Here, k is used to denote the order of the model and k ranges from 1 to the maximum order specified (M). The first coefficient is:

$$a_{1,1} = R_1 .$$

All of the $a_{k,k}$ coefficients are partial autocorrelation coefficients. The equation for all partial autocorrelation coefficients, $k = 2, \cdots, M$, is:

$$a_{k,k} = \frac{R_k - \sum\limits_{j=1}^{k-1} R_j a_{k-1,k-j}}{1 - \sum\limits_{j=1}^{k-1} R_j a_{k-1,j}} .$$

All of the remaining coefficients are calculated for k from 2 to the maximum order and $i=1, \cdots, k-1$:

$$a_{k,i} = a_{k-1,i} - a_{k-1,k-i} a_{k,k} .$$

For each order k from 1 to the maximum order, the residual variance is computed as:

$$RESIDVAR_k = \left[1 - \sum\limits_{j=1}^{k} R_j a_{k,j} \right] [Var\ X] .$$

The residual time series is computed for the maximum order (M). There are $(T-M)$ residual points. The equation used is:

$$E_t = x_t - \sum\limits_{j=1}^{M} a_{M,j} x_{t-j} ,$$

in which $t = M+1, \cdots, T$.

The autocorrelations of the residual (r_j) are computed using:

$$r_j = \frac{\sum\limits_{t=1}^{T-M-j} (E_t - \bar{E})(E_{t+j} - \bar{E})}{\sum\limits_{t=1}^{T-M} (E_t - \bar{E})^2} .$$

in which j was chosen by the authors to go from 1 to $(T-M)/3$.

The Box-Pierce statistic is next calculated as:

$$Q = (T-M) \sum\limits_{j=1}^{J} r_j^2 ,$$

in which $J = (T-M)/3$, as in the autocorrelations of the residuals.

The Box-Pierce statistic is distributed as chi-square with $(J-M)$ degrees of freedom.

For further significance testing, the inverse of the striped covariance matrix is computed. In the notation used here, the striped covariance matrix is defined as:

$$
A = \begin{bmatrix}
C_0 & C_1 & C_2 & C_3 & \cdots & C_{M-1} \\
C_1 & C_0 & C_1 & C_2 & \cdots & C_{M-2} \\
C_2 & C_1 & C_0 & C_1 & \cdots & C_{M-3} \\
\cdot & \cdot & \cdot & \cdot & & \cdot \\
\cdot & \cdot & \cdot & \cdot & & \cdot \\
\cdot & \cdot & \cdot & \cdot & & \cdot \\
C_{M-1} & C_{M-2} & C_{M-3} & C_{M-4} & \cdots & C_0
\end{bmatrix}
$$

Since this matrix is a positive definite symmetric matrix, the Bauer-Reinsch method of inversion could be used as described in Nash (1979).

The residual variance after the Mth step is:

$$
s^2 = \frac{1}{T-M} \sum_{t=1}^{T-M} (E_t - \bar{E})^2 ,
$$

which is used in the computation of the standard deviation of the final model coefficients. The standard deviation $(SD(a_{M,j}))$, with $j = 1, \cdots, M$, is the square root of the diagonal of:

$$
\frac{s^2}{T} A^{-1} ,
$$

in which A^{-1} is the inverse of the striped covariance matrix above.

Finally, Student's t-ratio can be computed as:

$$
t = a_{M,j} / SD(a_{M,j}) ,
$$

with degrees of freedom equal to $(T-M)$.

Program input

1. First input card.
   Cols. 2-5   Number of data points (T). Maximum is 1000.
   Cols. 7-10   Maximum order of the autoregressive model (M).
   Col. 20   Number of format cards to follow. Maximum

is 5.

Col. 25   Set to 1 if residual (error) series is to be punched.

2. Second input card.

Cols. 1-80   The title to be printed for identification of output.

3. Third input card or cards.

Cols. 1-80   Data format is specified on this card or cards. Standard FORTRAN F-format is used, enclosed in parentheses.

4. Data cards.

Cols. 1-80   Data cards follow the last format card. The data are punched in the format specified on the format card or cards.

Program output

When the program is run, the following information is output:

1. Program name: ARFIT PROGRAM.
2. The title for the printout identification.
3. Number of data points specified (T).
4. Maximum order of the autoregressive model specified (M).
5. Number of format cards specified.
6. Data format is repeated back to the user.
7. The first and last data points are printed for the user to check format.
8. Autocovariances of original data (C_j).
9. Autocorrelations of original data (R_j).
10. Initial variance of original data $(Var\ X)$.
11. A matrix of the residual variance, autoregressive model coefficients $(a_{k,i})$, and partial autocorrelation coefficients $(a_{k,k})$ for each order of the model up to the maximum specified.
12. The residual time series of the autoregressive fit (E_t). The residual time series is punched one to a card if requested on first input card.
13. Autocorrelations of the residual (r_j).
14. The Box-Pierce statistic and degrees of freedom.
15. A table of the coefficients of the final model $(a_{M,j})$ with the standard deviation for each and Student's t-ratio with degrees of freedom for significance testing.
16. The end of program message.

Special notes

1. If the number of data points is required to be greater than 1000, all array dimensions in the program will have to be increased and the four statements which check the maximum near the beginning of the program will have to be modified. The arrays X, IND and RES, should be set to a size equal to the number of data points. The arrays C, R, TRAT and SD were chosen to be one-sixth of the number of data points in size since the maximum order is usually less than or equal to one-sixth of the number of data points. The array A must be at least $(M(M+1)/2)$, in which M is the maximum order.
2. See 2, 3 and 4 under Special Notes for program DETRND.

Reference

Nash, J. C. (1979) *Compact numerical methods for computers: linear algebra and function minimisation.* New York: Halsted Press.

Example

Sample input

```
  60     3          1      1
SAMPLE  RUN  OF  AUTOREGRESSIVE  MODEL  FITTING  PROGRAM
(F4.2)
1.61
1.00
1.38
2.05
2.21
2.00
1.35
2.17
2.25
2.71
3.39
2.50
3.56
4.00
3.76
3.29
3.28
2.31
2.50
2.33
2.27
2.35
3.33
1.33
2.19
1.44
1.24
```

34

1.44
4.40
4.50
3.56
3.00
2.29
1.83
2.24
1.39
2.82
3.29
3.33
3.33
2.71
2.29
1.38
2.83
1.00
1.33
3.28
3.12
3.59
3.24
2.56
1.44
1.17
2.29
4.00
4.47
4.60
3.37
2.42
2.20

Sample output

ARFIT PROGRAM

SAMPLE RUN OF AUTOREGRESSIVE MODEL FITTING PROGRAM

60 DATA POINTS
MAXIMUM ORDER OF THE AUTOREGRESSIVE MODEL IS 3
1 FORMAT CARDS
DATA FORMAT: (F4.2)
RESIDUAL (ERROR) SERIES WILL GO TO PUNCH.   PUNCH FORMAT: (I5,F8.3)
FIRST DATA POINT IS   1.6100
LAST DATA POINT IS   2.2000

AUTOCOVARIANCES OF ORIGINAL DATA

ORDER	C	ORDER	C	ORDER	C	ORDER	C
1	.509	2	.159	3	-.139		

AUTOCORRELATIONS OF ORIGINAL DATA

ORDER	R	ORDER	R	ORDER	R	ORDER	R
1	.564	2	.176	3	-.154		

INITIAL VARIANCE OF ORIGINAL DATA =   .903

THE AUTOREGRESSIVE COEFFICIENTS,
PARTIAL AUTOCORRELATION COEFFICIENTS AND RESIDUAL VARIANCE

ORDER	RES. VAR.	A(1)	A(2)	A(3)	A(
1	.616	.564			
2	.589	.681	-.207		
3	.555	.631	-.044	-.240	

RESIDUAL TIME SERIES OF AR(3) FIT

TIME	ERROR	TIME	ERROR	TIME	ERROR	TIME	ERROR
4	1.610	5	1.218	6	1.027	7	.678
8	1.937	9	1.421	10	1.710	11	2.300
12	1.021	13	2.783	14	2.678	15	1.994
16	1.949	17	2.330	18	1.288	19	1.977
20	1.642	21	1.465	22	1.621	23	2.507
24	-.122	25	2.062	26	.917	27	.747
28	1.247	29	3.892	30	2.086	31	1.260
32	2.009	33	1.635	34	1.372	35	1.907
36	.607	37	2.481	38	2.110	39	1.712
40	2.051	41	1.546	42	1.526	43	.854
44	2.711	45	-.175	46	1.155	47	3.165
48	1.350	49	2.085	50	1.900	51	1.423
52	.830	53	1.152	54	2.230	55	2.953
56	2.328	57	2.506	58	1.625	59	1.570
60	1.926						

AUTOCORRELATIONS OF THE RESIDUAL

LAG J	R	LAG J	R	LAG J	R	LAG J	R
1	-.080	2	-.041	3	.101	4	-.141
5	-.219	6	.074	7	-.021	8	-.057
9	.082	10	-.020	11	-.235	12	.094
13	-.220	14	-.090	15	.176	16	-.095
17	.048	18	.224	19	.099		

BOX-PIERCE STATISTIC = 18.554, DISTRIBUTED AS CHI-SQUARE WITH DF = 16

STANDARD DEVIATION OF FINAL MODEL PARAMETER ESTIMATES
AND STUDENT'S T-RATIO WITH 57 DEGREES OF FREEDOM

	1	2	3
COEFFICIENTS	.631	-.044	-.240
SD	.124	.148	.124
T-RATIO	5.074	-.296	-1.932

END OF ARFIT PROGRAM

SPEC

Program description

The program SPEC performs a spectral analysis on a single time-series or calculates the periodogram for the series depending on the option specified. The procedures are described in Chapters 16 and 17 of Gottman's *Time-Series Analysis.*

Program steps

The program user supplies the time-series data (X), specifying the number of data points in the series (N) and the maximum lag $(MLAG)$ to be used in the spectral analysis computation. If the periodogram is to be calculated, $MLAG$ need not be specified as it is not used in the calculation.

If the number of data points specified is even, the last point is dropped from the series so that N becomes odd.

The mean of the data is then calculated and subtracted from each data point and the autocovariances (C) are calculated:

$$\bar{X} = \frac{1}{N}\sum_{t=1}^{N} x_t \ ,$$

$$C_0 = \frac{1}{N}\sum_{t=1}^{N} (x_t - \bar{X})^2$$

and

$$C_k = \frac{1}{N}\sum_{t=1}^{N-k} (x_t - \bar{X})(x_{t+k} - \bar{X}) \ .$$

For the spectral analysis, k goes from 1 to the maximum lag $(MLAG)$ specified. For the periodogram, k goes from 1 to N. C_0 is the variance of the data.

The autocorrelations (R) are each of the autocovariances divided by the variance of the data:

$$R_k = C_k/C_0 \ .$$

If the spectral analysis has been requested, the Tukey-Hanning window is employed. Weights are calculated for each lag k from 1 to $MLAG$:

$$W_k = \tfrac{1}{2}(1+\cos\pi\,(k/MLAG))\ .$$

If the periodogram calculation has been requested, all of the weights are set to 1.

The density estimate is calculated for each frequency in the overtone series:

$$F_j = j/N\ ,$$

in which $j = 1, 2, \cdots, (N-1)/2$. Then the density estimate (P) is :

$$P_{(F_j)} = \frac{1}{2\pi}\left\{C_0 + 2\sum_{k=1}^{MLAG} W_k C_k \cos(k2\pi F_j)\right\}$$

for each frequency F_j. The equation for the periodogram is similar:

$$P_{(F_j)} = \frac{1}{2\pi}\left\{C_0 + 2\sum_{k=1}^{N-1} C_k \cos(k2\pi F_j)\right\}$$

The weight, W, disappeared because it was set to 1.

The same calculation is also performed for $F = 0$ and $F = .5$.

The 95% confidence interval is calculated by:

$$\frac{EDF(P_{(F_j)})}{\chi^2_{EDF(.025)}} \leqslant P_{(F_j)} \leqslant \frac{EDF(P_{(F_j)})}{\chi^2_{EDF(.975)}}\ ,$$

and the 99% confidence interval is calculated by:

$$\frac{EDF(P_{(F_j)})}{\chi^2_{EDF(.005)}} \leqslant P_{(F_j)} \leqslant \frac{EDF(P_{(F_j)})}{\chi^2_{EDF(.995)}}\ ,$$

for each density estimate. For the spectral analysis we use the Tukey-Hanning window, for which:

$$EDF = \frac{8}{3}\frac{N}{MLAG}$$

for all $P_{(F_j)}$ except when $F = 0$ or $F = .5$,

$$EDF = EDF/2\ .$$

Reference for the calculation of the degrees of freedom is Anderson, 1971, page 531. For the periodogram, EDF is 2 and 1 when $F = 0$ or $F = .5$.

If the degrees of freedom are 30 or less, the program completes the above calculations for the confidence intervals. If the degrees of freedom are greater than 30, the program user will have to do the calculations.

The white noise spectrum value, which is the variance of the data divided by 2π, is shown on the plotted output of density estimates and confidence interval. Density estimates which are significantly different from white noise are clearly indicated on the plot where the line of pluses at the white noise spectrum value does not fall within the confidence interval indicated by asterisks. The top or bottom of the confidence interval may often be off of the plot.

Program input

1. First input card.
   Cols. 2-5   Number of data points (N) in the time-series. Maximum is 600.

   Cols. 8-10   Maximum number of lags to be used in the spectral analysis. Maximum is 599.

   Col. 15   Set to 1 if periodogram is wanted instead of spectral density estimates.

   Col. 20   Number of data format cards. Maximum is 5.

   Col. 25   Set to 1 if the density estimates or periodogram output is to go to punch.

2. Second input card.
   Cols. 1-80   The title to be printed for identification of output.

3. Next card or cards.
   Cols. 1-80   Time-series data format is specified on this card or cards. Standard FORTRAN F-format is used, enclosed in parentheses.

4. Time-series data cards.
   Cols. 1-80   The time-series data cards follow the last format card. The data are punched according to the format specified on the format card or cards.

Program output

When the program is run, the following information is output:
1. Program name: SPEC PROGRAM.
2. The title for the printout for later identification.
3. Specified number of data points in the time-series (N).
4. Specified number of lags ($MLAG$) to be used in the calculations.

5. The number of format cards specified.
6. The data format is printed back for the user.
7. The user is informed as to whether spectral densities of periodogram will be calculated.
8. If output is to go to punch, the punch format will be printed.
9. First and last data points are printed so that the user can check the input format.
10. The mean of the data is printed.
11. The variance of the data is printed.
12. The white noise spectrum value is printed.
13. Table of autocovariances (C).
14. Table of autocorrelations (R).
15. Spectral density estimates or periodogram with a value for each frequency F, degrees of freedom, and the 95% and 99% confidence intervals.
16. The plot of density estimates for each frequency, F. The density estimates are indicated by the symbol X. The 95% confidence interval is indicated by an asterisk on either side of the X unless the confidence interval value falls outside the range of the plot. The white noise spectrum value is indicated across the plot by a line of pluses.
17. The end of program message.

Special notes

1. If more than 600 data points are required, the dimension of arrays X and W must be changed to the required number of data points; dimension of arrays C and R must be the number of data points plus one; and the dimension of P should be half the number of data points. Statements near the start of the program which check for these maxima must be modified.

 A computer programmer should be able to make the χ^2 table (CHITAB) larger for degrees of freedom greater than 30 if desired. The table is in the order: .995 for the 1-30 degrees of freedom, then .005 for the 1-30 degrees of freedom, then .975 and .025. There are several statements in the program which check for the 30 maximum degrees of freedom which would also have to be modified.
2. See Special Notes 2, 3 and 4 for the DETRND program.

Reference

Anderson, T. W. (1971) *The statistical analysis of time-series.* New York: Wiley.

Example

Sample input

```
    60    10              1       1
SAMPLE RUN OF SPECTRAL ANALYSIS PROGRAM
(F4.2)
1.61
1.00
1.38
2.05
2.21
2.00
1.35
2.17
2.25
2.71
3.39
2.50
3.56
4.00
3.76
3.29
3.28
2.31
2.50
2.33
2.27
2.35
3.33
1.33
2.19
1.44
1.24
1.44
4.40
4.50
3.56
3.00
2.29
1.83
2.24
1.39
2.82
3.29
3.33
3.33
2.71
2.29
1.38
2.83
1.00
1.33
3.28
3.12
3.59
3.24
2.56
```

1.44
1.17
2.29
4.00
4.47
4.60
3.37
2.42
2.20

Sample output

SPEC PROGRAM

SAMPLE RUN OF SPECTRAL ANALYSIS PROGRAM

60 DATA POINTS
NUMBER OF LAGS TO BE USED IS 10
 1 FORMAT CARDS
DATA FORMAT: (F4.2)

SPECTRAL DENSITIES WILL BE CALCULATED

OUTPUT WILL GO TO PUNCH
PUNCH FORMAT IS (F10.4)

FIRST DATA POINT IS   1.610
LAST DATA POINT IS   2.200

THE MEAN OF THE DATA IS   2.582
THE VARIANCE OF THE DATA IS   .916
WHITE NOISE SPECTRUM VALUE (VARIANCE OVER 2 PI) IS   .146

TABLE OF AUTOCOVARIANCES

LAG	C	LAG	C	LAG	C	LAG	C	LAG	C
0	.916	1	.516	2	.167	3	-.128	4	-.290
5	-.260	6	-.055	7	.063	8	.125	9	.108
10	-.025								

TABLE OF AUTOCORRELATIONS

LAG	R	LAG	R	LAG	R	LAG	R	LAG	R
0	1.000	1	.564	2	.182	3	-.140	4	-.316
5	-.284	6	-.060	7	.069	8	.136	9	.118
10	-.028								

SPECTRAL DENSITY ESTIMATES

			CHI-SQUARE CONFIDENCE INTERVALS			
FREQ	DENSITY	D.F.	.05 LEVEL		.01 LEVEL	
0.000	.2225	8	.1015 --	.8168	.0811 --	1.3242
.017	.2318	16	.1286 --	.5368	.1082 --	.7211
.034	.2581	16	.1431 --	.5977	.1205 --	.8030
.051	.2970	16	.1647 --	.6880	.1387 --	.9242
.068	.3406	16	.1889 --	.7889	.1590 --	1.0597
.085	.3777	16	.2095 --	.8749	.1764 --	1.1753
.102	.3968	16	.2201 --	.9191	.1853 --	1.2346
.119	.3893	16	.2159 --	.9016	.1818 --	1.2112
.136	.3532	16	.1959 --	.8182	.1649 --	1.0991
.153	.2947	16	.1634 --	.6825	.1376 --	.9168
.169	.2253	16	.1250 --	.5219	.1052 --	.7011
.186	.1587	16	.0880 --	.3675	.0741 --	.4937
.203	.1053	16	.0584 --	.2440	.0492 --	.3277
.220	.0702	16	.0389 --	.1626	.0328 --	.2184
.237	.0523	16	.0290 --	.1210	.0244 --	.1626

.254	16	.0467	.1082	.0259 --	.0218 --	.1454
.271	16	.0477	.1105	.0265 --	.0223 --	.1485
.288	16	.0505	.1170	.0280 --	.0236 --	.1572
.305	16	.0524	.1213	.0290 --	.0245 --	.1629
.322	16	.0521	.1208	.0289 --	.0243 --	.1622
.339	16	.0498	.1154	.0276 --	.0233 --	.1551
.356	16	.0460	.1065	.0255 --	.0215 --	.1430
.373	16	.0413	.0957	.0229 --	.0193 --	.1286
.390	16	.0370	.0856	.0205 --	.0173 --	.1150
.407	16	.0338	.0784	.0188 --	.0158 --	.1053
.424	16	.0327	.0757	.0181 --	.0153 --	.1016
.441	16	.0334	.0775	.0186 --	.0156 --	.1041
.458	16	.0356	.0824	.0197 --	.0166 --	.1106
.475	16	.0379	.0878	.0210 --	.0177 --	.1180
.492	16	.0394	.0914	.0219 --	.0184 --	.1227
.500	8	.0396	.1455	.0181 --	.0144 --	.2359

PLOT OF DENSITY ESTIMATES
DENSITY VALUES APPEAR ACROSS TOP AND FREQUENCIES DOWN
THE LEFT SIDE OF PLOT.
THE DENSITY IS INDICATED BY X AND CONFIDENCE INTERVAL
AT .05 LEVEL IS INDICATED BY ASTERISKS.
THE WHITE NOISE SPECTRUM VALUE IS SHOWN BY THE
LINE OF PLUSES.

```
        .0327    .0847    .1367    .1887    .2407    .2928    .3448    .396
       +········+········+········+········+········+········+········+
0.000  +
.017   +              .           X
.034   +                .   X
.051   +                  .              X
.068   +                .                           X
.085   +              .                                     X
.102   +              .                                              X
```

```
.119           X
.136          X
.153     X
.169           .
.186      X
.203      X   .
.220  X   X +  .
.237      X
.254 X
.271 X
.288 X
.305 X
.322 X
.339 X
.356 X
.373 X
.390X
.407X
.424X
.441X
.458X
.475X
.492 X
.500 X
```

END OF SPEC PROGRAM

Package 2.
Applications of
univariate model fitting

FORCST

Program description

The program FORCST is the best one-step ahead prediction program using the autoregressive model fitting process. The procedure is defined in Chapter 21 of Gottman's *Time-Series Analysis.*

Program steps

The user supplies the time-series (X) with N data points and the maximum order of the autoregressive model, *MORD*, as described for the program ARFIT in this manual.

The autocovariances (C), autocorrelations (R), and autocorrelation coefficients (a) are computed as described for the ARFIT program.

Then the best one-step ahead prediction is:

$$\hat{x}_{N+1} = \sum_{s=1}^{MORD} a_s X_{N-s+1} .$$

The one-step ahead prediction variance (see Chapter 21; set $s = 1$) is:

$$\hat{\sigma}_e^2 = \hat{\sigma}_x^2 (1 - \hat{a}_1 r_1 - \hat{a}_2 r_2 - \cdots \hat{a}_{MORD} r_{MORD}) ,$$

in which $\hat{\sigma}_x^2$ is the variance estimate of data to the point of prediction:

$$\hat{\sigma}_x^2 = \frac{1}{N} \sum_{t=1}^{N} (x_t - \overline{X})^2 .$$

Note that in the program N (instead of $N-1$) is used as the divisor in the equation above to agree with the procedure used in the program ARFIT.

The one-step ahead prediction variance is used to compute the confidence intervals for the predictions at the .05 and .01 levels:

$$\hat{x}_{N+1} \pm 1.96 \hat{\sigma}_e^2$$

and

$$\hat{x}_{N+1} \pm 2.58 \hat{\sigma}_e^2 .$$

51

Program input

1. First input card.
   Cols. 2-5   Number of data points (N) in the time-series. Maximum is 1000.
   Cols. 7-10   Maximum order of the autoregressive model.
   Col. 20   Number of format cards to follow. Maximum is 5.
2. Second input card.
   Cols. 1-80   The title to be printed on output for identification.
3. Next card or cards.
   Cols. 1-80   Time-series data format is specified on this card or cards. Standard FORTRAN F-format is used, enclosed in parentheses.
4. Time-series data cards.
   Cols. 1-80   The time-series data cards follow the last format card. The data are punched according to the format specified on the format card or cards.

Program output

When the program is run, the following information is output:
1. Program name: FORCST PROGRAM.
2. The title for the printout identification.
3. Number of data points (N) specified.
4. Maximum order of the autoregressive model $(MORD)$ specified.
5. Number of data format cards specified.
6. The data format is printed back for the user.
7. The first and last data points are printed for the user to check format.
8. Variance estimate of data up to point of prediction $(\hat{\sigma}_x^2)$.
9. The best one-step prediction $(\hat{x}_{N+1})$.
10. The one-step ahead prediction variance $(\hat{\sigma}_e^2)$.
11. The confidence interval for the prediction at the .05 level.
12. The confidence interval for the prediction at the .01 level.
13. End of the program message.

Special notes

1. If the number of data points in the time-series is required to be larger than 1000, the sizes of arrays in the program will have to be changed and the statements near the beginning of the program which check this maximum will have to be modified. The

size of array X should be set to the number of data points in the time-series. The arrays C and R are set at one-sixth of the number of data points in the time-series or the maximum order of the autoregressive model. The array A must be at least $(M(M+1)/2)$ in which M is the maximum order.

2. See Special Notes 3 and 4 for the program DETRND.

Example

Sample input

```
    48       2             1
SAMPLE RUN OF ONE-STEP AHEAD PREDICTION PROGRAM
(F4.2)
1.61
1.00
1.38
2.05
2.21
2.00
1.35
2.17
2.25
2.71
3.39
2.50
3.56
4.00
3.76
3.29
3.28
2.31
2.50
2.33
2.27
2.35
3.33
1.33
2.19
1.44
1.24
1.44
4.40
4.50
3.56
3.00
2.29
1.83
2.24
1.39
2.82
3.29
3.33
3.33
2.71
2.29
```

54

1.38
2.83
1.00
1.33
3.28
3.12

Sample output

FORCST PROGRAM

SAMPLE RUN OF ONE-STEP AHEAD PREDICTION PROGRAM

    48 DATA POINTS
    MAXIMUM ORDER OF THE AUTOREGRESSIVE MODEL IS   2
    1 FORMAT CARDS
    DATA FORMAT: (F4.2)
    THE FIRST DATA POINT IS   1.6100
    THE LAST DATA POINT IS   3.1200

VARIANCE ESTIMATE OF DATA UP TO POINT OF PREDICTION IS: .800

THE BEST ONE-STEP PREDICTION FOR X(48+1) IS: 1.426

THE ONE-STEP AHEAD PREDICTION VARIANCE IS: .477

FOR CONFIDENCE INTERVAL WITH PROBABILITY OF (1-.05)
    THE PREDICTION LIES BETWEEN   .492 AND   2.361
FOR CONFIDENCE INTERVAL WITH PROBABILITY OF (1-.01)
    THE PREDICTION LIES BETWEEN   .196 AND   2.656

END OF FORCST PROGRAM

ITSE

Program description

The program ITSE performs the interrupted time-series analysis using the linear autoregressive model as described in Chapter 26 of Gottman's *Time-Series Analysis.*

Program steps

The program user supplies the time-series data **Y**, indicating the number of data points before (NB) and after (NA) intervention. The user also supplies the order ($NORD$) of the autoregressive model.

The data before intervention are represented as:

$$y_t = m_1 t + b_1 + \sum_{i=1}^{NORD} a_i y_{t-i} + e_t \, ,$$

in which $t = NORD+1, NORD+2, \cdots, NB$. The data after intervention are represented as:

$$y_t = m_2 t + b_2 + \sum_{i=1}^{NORD} a_i y_{t-i} + e_t \, ,$$

in which $t = NB+1, NB+2, \cdots, NB+NA$. In matrix form, these equations become:

$$\mathbf{Y} = \mathbf{X}\boldsymbol{\beta} + \mathbf{E} \, .$$

The matrix **X** is set up as:

$$\mathbf{X} = \begin{bmatrix} 1 & 1 & 0 & 0 & y_{NORD} & y_{NORD-1} & \cdots & y_1 \\ 1 & 2 & 0 & 0 & y_{NORD+1} & y_{NORD} & \cdots & y_2 \\ \vdots & \vdots & \vdots & \vdots & \vdots & \vdots & & \vdots \\ 1 & NB & 0 & 0 & y_{NB-1} & y_{NB-2} & \cdots & y_{NB-NORD} \\ 0 & 0 & 1 & 1 & y_{NB} & y_{NB-1} & \cdots & y_{NB-NORD+1} \\ 0 & 0 & 1 & 2 & y_{NB+1} & y_{NB} & \cdots & y_{NB-NORD+2} \\ \vdots & \vdots & \vdots & \vdots & \vdots & \vdots & & \vdots \\ 0 & 0 & 1 & NA & y_{NB+NA-1} & y_{NB+NA-2} & \cdots & y_{NB+NA-NORD} \end{bmatrix} \, .$$

Y is the column vector of data

$$\mathbf{Y} = \begin{bmatrix} y_{NORD+1} \\ y_{NORD+2} \\ \vdots \\ y_{NB} \\ y_{NB+1} \\ y_{NB+2} \\ \vdots \\ y_{NB+NA} \end{bmatrix}.$$

$\hat{\beta}$ contains the estimates of the parameters upon the completion of the least squares solution:

$$\hat{\beta} = (\mathbf{X}^T\mathbf{X})^{-1}\mathbf{X}^T\mathbf{Y}.$$

In this computation, the Bauer-Reinsch method of matrix inversion is performed as described in Nash (1979).

$$\beta = \begin{bmatrix} b_1 \\ m_1 \\ b_2 \\ m_2 \\ a_1 \\ a_2 \\ \vdots \\ a_{NORD} \end{bmatrix}.$$

The elements of β are used in the computation of B_1, B_2, M_1 and M_2, which can be used to show the trend lines of the data:

$$M = \frac{m}{1 - \sum a_i}$$

and

$$B = \frac{b - M \sum i a_i}{1 - \sum a_i}.$$

The sum of squares for error (SSE) is computed:

$$SSE = \frac{1}{NB+NA-2(NORD)-4}(\mathbf{Y}-\mathbf{X}\hat{\boldsymbol{\beta}})^T(\mathbf{Y}-\mathbf{X}\hat{\boldsymbol{\beta}})$$

and

$$SE = \sqrt{SSE} \ .$$

The t-test is performed on each element of $\hat{\boldsymbol{\beta}}$:

$$t_i = \frac{\hat{\beta}_i}{SE \sqrt{i\text{th diagonal element of}(\mathbf{X}^T\mathbf{X})^{-1}}}$$

with $NB+NA-2(NORD)-4$ degrees of freedom.
The t-test which tests if b_1 equals b_2 is:

$$t = \frac{b_1-b_2}{\sqrt{SSE/(NB+NA-2(NORD)-4)}} \ ,$$

and the t-test which tests if m_1 equals m_2 is:

$$t = \frac{m_1-m_2}{\sqrt{SSE/(NB+NA-2(NORD)-4)}} \ .$$

Test the .05 level of significance in both cases at .025 with $(NB+NA-2(NORD)-4)$ degrees of freedom.

If the sum of the autoregressive parameters (a) in β is greater than .7, the program will print a message. Subtracting the first lag of the data from the data and inputting this difference to the ITSE program may take care of the problem. That is:

$$y'_t = y_t - y_{t-1} \ .$$

For the F-test, a similar computation must be performed on the data with the reduced model:

$$y_t = mt+b+ \sum_{i=1}^{NORD} a_i y_{t-i}+e_t \ ,$$

in which all the data are considered together as though there were no intervention. In this case, $\mathbf{X}$ is set up as:

$$X = \begin{bmatrix} 1 & NORD+1 & y_{NORD} & y_{NORD-1} & \cdots & y_1 \\ 1 & NORD+2 & y_{NORD+1} & y_{NORD} & \cdots & y_2 \\ 1 & NORD+3 & y_{NORD+2} & y_{NORD+1} & \cdots & y_3 \\ \vdots & \vdots & \vdots & \vdots & & \vdots \\ 1 & NB+NA & y_{NB+NA-1} & y_{NB+NA-2} & \cdots & y_{NB+NA-NORD} \end{bmatrix}.$$

This time the result of the least squares solution is:

$$\beta = \begin{bmatrix} b_1 \\ m_1 \\ a_1 \\ a_2 \\ \vdots \\ a_{NORD} \end{bmatrix}.$$

The calculations of M, B and SSE are the same for the reduced model as for the full model.

For the F-test, the SSE for the full model is denoted SS_1 and the SSE for the reduced model is denoted SS_0.

$$F = \frac{(SS_0 - SS_1)/2}{SS_1/(NB+NA-2(NORD)-4)}$$

with 2 and $NB+NA-2(NORD)-4$ degrees of freedom.

Program input

1. First input card.
   - Cols. 2-5   Number of data points (NB) in the time-series before intervention.
   - Cols. 7-10   Number of data points (NA) in the time-series after intervention. $NB+NA-NORD$ must not be greater than 400.
   - Cols. 12-15   The order of the autoregressive model ($NORD$). Maximum is 50.
   - Col. 20   Number of data format cards. Maximum is 5.
2. The second input card.
   - Cols. 1-80   The title to be printed on the output.
3. Next card or cards.
   - Cols. 1-80   Time-series data format is specified on this card or cards. Standard FORTRAN F-format is

used, enclosed in parentheses.

4. Time-series data cards.

    Cols. 1-80   The time-series data cards follow the last format card. The data are punched according to the format specified on the format card or cards.

Program output

When the program is run, the following information is output:

1. Program name: ITSE PROGRAM.
2. The title for the printout identification.
3. Specified number of data points in time-series before intervention (NB).
4. Specified number of data points in time-series after intervention (NA).
5. The order of the autoregressive model specified ($NORD$).
6. The number of format cards specified.
7. The data format is printed back for the user.
8. The first and last data points are printed for the user to check format.
9. b_1, m_1, B_1, M_1, b_2, m_2, B_2 and M_2 for the full model are printed.
10. The elements of BETA and t-test of these elements are output with degrees of freedom.
11. b, m, B and M of the reduced model.
12. The F-test with degrees of freedom.
13. End of program message.

Special notes

1. If more data points are required in the time-series or if the autoregressive order is larger than 50, array dimensions and statements which check for these maxima must be modified in the program. Y, WORKA, WORKB and one dimension of X, XT and TEMP must be at least equal to the total number of data points. BETA and the other dimension of X, XT and TEMP must be at least as large as the autoregressive order plus 4. XINV size must be at least $(M(M+1)/2)$, in which M is the autoregressive model order.
2. See Special Notes 3 and 4 for the DETRND program.

62

Reference

Nash, J. C. (1979) *Compact numerical methods for computers: linear algebra and function minimisation.* New York: Halsted Press.

Example

Sample input

```
    46     47     2     1
SAMPLE RUN OF INTERRUPTED TIME SERIES ANALYSIS PROGRAM
(F10.1)
            59.0          1.0          0.0        1879.0
            72.0          1.0          0.0        1880.0
            67.2          1.0          0.0        1881.0
            60.5          1.0          0.0        1882.0
            60.7          1.0          0.0        1883.0
            69.7          1.0          0.0        1884.0
            61.2          1.0          0.0        1885.0
            63.3          1.0          0.0        1886.0
            60.0          1.0          0.0        1887.0
            67.5          1.0          0.0        1888.0
            63.0          1.0          0.0        1889.0
            59.2          1.0          0.0        1890.0
            59.6          1.0          0.0        1891.0
            57.7          1.0          0.0        1892.0
            57.3          1.0          0.0        1893.0
            58.9          1.0          0.0        1894.0
            59.8          1.0          0.0        1895.0
            57.0          1.0          0.0        1896.0
            62.4          1.0          0.0        1897.0
            62.1          1.0          0.0        1898.0
            68.3          1.0          0.0        1899.0
            69.8          1.0          0.0        1900.0
            65.7          1.0          0.0        1901.0
            58.9          1.0          0.0        1902.0
            62.1          1.0          0.0        1903.0
            63.3          1.0          0.0        1904.0
            68.9          1.0          0.0        1905.0
            63.0          1.0          0.0        1906.0
            57.5          1.0          0.0        1907.0
            69.2          1.0          0.0        1908.0
            55.3          1.0          0.0        1909.0
            53.7          1.0          0.0        1910.0
            54.3          1.0          0.0        1911.0
            55.7          1.0          0.0        1912.0
            55.6          1.0          0.0        1913.0
            59.4          1.0          0.0        1914.0
            62.4          1.0          0.0        1915.0
            57.7          1.0          0.0        1916.0
            60.1          1.0          0.0        1917.0
            55.7          1.0          0.0        1918.0
            50.8          1.0          0.0        1919.0
            49.5          1.0          0.0        1920.0
            47.6          1.0          0.0        1921.0
            49.4          1.0          0.0        1922.0
            58.9          1.0          0.0        1923.0
```

53.4	1.0	0.0	1924.0
54.0	0.0	1.0	1925.0
57.6	0.0	1.0	1926.0
56.2	0.0	1.0	1927.0
68.6	0.0	1.0	1928.0
73.8	0.0	1.0	1929.0
69.7	0.0	1.0	1930.0
74.6	0.0	1.0	1931.0
74.2	0.0	1.0	1932.0
74.4	0.0	1.0	1933.0
79.2	0.0	1.0	1934.0
76.3	0.0	1.0	1935.0
78.3	0.0	1.0	1936.0
80.8	0.0	1.0	1937.0
77.3	0.0	1.0	1938.0
79.3	0.0	1.0	1939.0
84.4	0.0	1.0	1940.0
82.0	0.0	1.0	1941.0
79.8	0.0	1.0	1942.0
78.8	0.0	1.0	1943.0
78.0	0.0	1.0	1944.0
82.6	0.0	1.0	1945.0
81.2	0.0	1.0	1946.0
83.7	0.0	1.0	1947.0
82.6	0.0	1.0	1948.0
81.9	0.0	1.0	1949.0
85.6	0.0	1.0	1950.0
84.3	0.0	1.0	1951.0
85.7	0.0	1.0	1952.0
86.0	0.0	1.0	1953.0
83.6	0.0	1.0	1954.0
85.0	0.0	1.0	1955.0
83.2	0.0	1.0	1956.0
82.8	0.0	1.0	1957.0
83.2	0.0	1.0	1958.0
85.0	0.0	1.0	1959.0
88.7	0.0	1.0	1960.0
88.3	0.0	1.0	1961.0
88.7	0.0	1.0	1962.0
89.3	0.0	1.0	1963.0
90.5	0.0	1.0	1964.0
89.0	0.0	1.0	1965.0
90.0	0.0	1.0	1966.0
88.2	0.0	1.0	1967.0
89.5	0.0	1.0	1968.0
80.0	0.0	1.0	1969.0
82.5	0.0	1.0	1970.0
77.0	0.0	1.0	1971.0

64

Sample output

ITSE PROGRAM

SAMPLE RUN OF INTERRUPTED TIME SERIES ANALYSIS PROGRAM

```
46 DATA POINTS BEFORE INTERVENTION
47 DATA POINTS AFTER INTERVENTION
ORDER OF THE AUTOREGRESSIVE MODEL IS      2
 1 FORMAT CARDS
DATA FORMAT: (F10.1)
THE FIRST DATA POINT IS    59.0000
THE LAST DATA POINT IS    77.0000
```

```
              THE FULL MODEL
SMALL B1 =    19.210      SMALL M1 =    -.063
LARGE B1 =    66.080      LARGE M1 =    -.209

SMALL B2 =    22.871      SMALL M2 =     .077
LARGE B2 =    72.619      LARGE M2 =     .253
```

```
              T-TEST ON ELEMENTS OF BETA
                   85 DEGREES OF FREEDOM
   BETA(  1)=    19.210        T=    3.111
   BETA(  2)=     -.063        T=   -1.223
   BETA(  3)=    22.871        T=    3.714
   BETA(  4)=      .077        T=    1.141
   BETA(  5)=      .537        T=    5.270
   BETA(  6)=      .161        T=    1.571
```

```
T-TEST FOR SMALL B1 = SMALL B2:  T =  -8.727     DF =   85
T-TEST FOR SMALL M1 = SMALL M2:  T =   -.333     DF =   85
```

BE CAREFUL THE SUM OF THE AUTOREGRESSIVE PARAMETERS
IS GREATER THAN .7 -- SEE CHAP. 26 OF GOTTMAN.

```
              THE REDUCED MODEL
SMALL B =    7.182      SMALL M =    .058
LARGE B =   43.154      LARGE M =    .412
```

```
              F-TEST
 F=    5.611 WITH (2,  85) DEGREES OF FREEDOM
```

END OF ITSE PROGRAM

Package 3.
Multivariate time-series analysis

BIVAR

Program description

The program BIVAR is the time-domain bivariate analysis program. The program performs the analysis described as the procedure of Gottman and Ringland in Chapter 25 of Gottman's *Time-Series Analysis.*

Program steps

The program user supplies the initial value for the A, B, C and D of the model and the critical value (CR) for the testing of the t-statistic. (Spectral analysis may be used to aid in the determination of the initial value of A, B, C and D.) If no critical value is specified by the user, 1.6 is used. Two corresponding time-series are also supplied by the user. N is the number of data points in each time-series.

In the first pass through the program, the main series is the first time-series $(Y1_i)$ supplied by the user and the explanatory series is the second time-series $(Y2_i)$. Lags of each series are determined according to the initial value of A and B, referred to as INT in the program.

The main series is defined as:

$$Y_i = Y1_{(i+INT)},$$

in which $i = 1, 2, \cdots, T$ and $T = N - INT$.

The matrix of the lagged main series is:

$$X1_{(i,j)} = Y1_{(INT+i-j)},$$

and the lagged explanatory series is:

$$X2_{(i,j)} = Y2_{(INT+i-j)},$$

in which $i = 1, 2, \cdots, T$ and $j = 1, 2, \cdots, INT$ in both cases.

For each fitting of the model, the matrix $\mathbf{X}$ is constructed as:

$$\mathbf{X} = \begin{bmatrix} X1_{1,1} & \cdots & X1_{1,A} & X2_{1,1} & \cdots & X2_{1,B} \\ X1_{2,1} & \cdots & X1_{2,A} & X2_{2,1} & \cdots & X2_{2,B} \\ \vdots & & \vdots & \vdots & & \vdots \\ X1_{T,1} & \cdots & X1_{T,A} & X2_{T,1} & \cdots & X2_{T,B} \end{bmatrix}.$$

Either A or B may be zero.

The least-square parameter estimates (Θ_i) are computed by the matrix equation:

$$\Theta_i = (\mathbf{X}^T\mathbf{X})^{-1}\mathbf{X}^T\mathbf{Y} \; ,$$

in which $i = 1, 2, \cdots, A+B$. The matrix inversion for this equation is performed by the subroutine based on the Bauer-Reinsch method of inversion as described in Nash (1979). In the program, the array BETA contains the computed values of theta.

The sum of squares for error (SSE) is calculated by the matrix equation:

$$SSE = (\mathbf{Y}-\mathbf{X}\Theta)^T(\mathbf{Y}-\mathbf{X}\Theta) \; ,$$

and the mean square error term is:

$$MSE = SSE/(T-A-B) \; .$$

The diagonal of the variance-covariance matrix is:

$$SE_i = \sqrt{[i\text{th diagonal element in } (\mathbf{X}^T\mathbf{X})^{-1}][MSE]} \; .$$

These values are used in the calculation of the t-statistic:

$$TT_i = \Theta_i/SE_i \; .$$

In both equations $i = 1, 2, \cdots, A+B$.

The likelihood ratio is computed as:

$$LR = T\ln(SSE/T) \; .$$

This completes what is defined as fitting the model.

At the end of this procedure, if both A and B are not zero, the t-statistic (TT_i) is tested against the critical value (CR). If the absolute value of TT_A is less than CR, A is reduced by 1. If the absolute value of $TT_{(A+B)}$ is less than CR, B is reduced by 1. If neither of these are less than the critical value, the best model has been found to describe the series. In the case that the best model has not been found and either A or B is not equal to zero, the model is again fit for the new A and B pair. In the case that the best model has not been found and A and B are both zero, one is added to either A or B corresponding to whether the absolute value of TT_A is larger or the absolute value of $TT_{(A+B)}$ is larger and then the model is fit for the new A and B pair.

If, at the end of the model fitting procedure, either A or B is zero, the absolute value of the t-statistic corresponding to the non-zero value (either A or B) is tested against the critical value. If it is greater than the critical value, the best model has been found to describe the series. If it is less than the critical value, one is subtracted from the non-zero value of either A or B. If that value reaches zero also, the sum of squares for error term becomes:

$$SSE = \sum_{i=1}^{T} Y_i^2 \, ,$$

and the process is considered completed as if the best model had been found. If the new value is not zero, the model is fit for the new A and B pair.

When the best model has been found to describe the series, the summary table is printed for the user.

In the summary table there are five lines. In the first line, A and B are both equal to the initial value (INT), the sum of squares for error and likelihood ratio are as determined by the first fit of the model. The second line is the best model fit and the sum of squares for error and likelihood ratio terms are as they were determined by that last pass through the model fitting and testing procedure.

For the third line of the summary table, one pass is made through the model fitting procedure with A equal to its final value and B equal to zero. If A is not equal to zero, the SSE term and likelihood ratio term are as calculated in that pass. If A is zero, these two terms will be the same as for line 5 of the summary table.

For the fourth line, one pass is made through the model fitting procedure with A equal to the initial value (INT) and B equal to zero. The SSE and likelihood ratio terms are as calculated in that procedure.

In the fifth line, in which A and B are both zero,

$$SSE = \sum_{i=1}^{T} Y_i^2$$

and the likelihood ratio is determined, using SSE term just calculated:

$$LR = T \ln(SSE/T) \, .$$

The Q-test statistic is calculated for lines 1 versus 2 as likelihood ratio for line 2 minus the likelihood ratio for line 1 and it is distributed as chi-square with $(2INT-A-B)$ degrees of freedom.

The Q-test statistic for comparison of lines 2 versus 3 is likelihood ratio for line 3 minus likelihood ratio for line 2 and it is distributed as chi-square with B degrees of freedom.

Finally, the Q-test statistic for line 3 versus 4 is likelihood ratio for line 3 minus the likelihood ratio for line 4. It is distributed as chi-square with $(INT-A)$ degrees of freedom.

For further significance testing:

$$z = \frac{(Q/DF)-1}{\sqrt{2/DF}} \qquad \sim N(0,1) \, .$$

For large DF, compare z with a unit normal distribution. (For example, at alpha = .05, z should be greater than 1.96 for significance.)

If the message "INVERSE CAN'T BE PERFORMED, MATRIX IS SINGULAR" is printed instead of the summary table, it may be because one time-series is constant.

From this first summary table, it is determined if the second series predicts the first.

In the second pass through the program, the second series $(Y2_i)$ becomes the main series (Y_i) and the first series $(Y1_i)$ becomes the explanatory series. Although the calculation then deal with C and D instead of A and B, the names A and B are still used within the program. The printout will be appropriately labelled, however.

The second summary table is used to determine if the first series predicts the second.

Program input

1. First input card.

    Cols. 3-5   Number of data points (N) in each time series. Maximum is 400.

    Cols. 8-10   The initial value (INT) for A, B, C and D of the model. Spectral analysis may be used in the determination of this value.

    Cols. 11-15   The critical value (CR) for testing the t-statistic. If not specified, 1.6 is used.

    Col. 18   Set to 1 if data for the two series alternate on input (e.g., $Y1_1$, $Y2_1$, $Y1_2$, $Y2_2$, etc.). Otherwise all data for the first series $(Y1)$ will be read followed by all data for the second series $(Y2)$ with the second series beginning on a new card if there is more than one data point on a card.

    Col. 20   Number of format cards to follow. Maximum is 5.

    Col. 30   Set to 1 if only summary tables are to be printed.

2. Second input card.

    Cols. 1-80   The title which the user wishes to have printed to label the output.

3. Next input card or cards.

    Cols. 1-80   Data format is specified on this card or cards. Standard FORTRAN F-format is used, enclosed in parentheses.

4. Time-series data cards.

    Cols. 1-80   Data cards for the time-series follow last format card. Data are punched in the format specified on the format card or cards and as specified by column 18 of the first input card.

Program output

When the program is run, the following information is output:
1. Program name: BIVAR PROGRAM.
2. The title for the printout identification.
3. Number of data points (N) specified.
4. Initial value (INT) of A, B, C and D specified.
5. Critical value (CR) to be used.
6. Number of format cards specified.
7. Data format is repeated back to the user.
8. The first and last data points in each series are printed for the user to check format.
9. Number of data points (T) to be used in the computations after the lags are taken.
10. After each fit of the model is completed, the current value of A and B are printed with the sum of squares for error (SSE), the mean square error (MSE), and the likelihood ratio. At this time, a table is also printed showing the least-square parameter estimate and the t-statistic for each lag of each series.
11. Finally, a summary table is printed showing again the number of data points (T) used from each series. For each model used in the likelihood procedure, A and B are shown along with the sum of squares for error and the likelihood ratio. At the end of the summary table, the Q-statistic with degrees of freedom are shown for each pair of models being compared.
12. The whole output is repeated with the two series interchanged.

Special notes

1. If the number of data points (N) in each series is required to be larger than 400, the dimension in arrays Y1, Y2, Y, WORKA and WORKB will have to be increased to be equal to the number of data points and the program statements near the start of the program which check for the maximum value will have to be modified. One dimension of each of the arrays X, XT, and TEMP will also have to be changed to one less than the number of data points and the other dimension, as well as the single dimension of BETA, must be at least twice as large as the initial value of A, B, C and D. The dimension of XINV must be larger than ½ the initial value of A, B, C and D times that initial value plus one.
2. See Special Notes 3 and 4 of the DETRND program.

72

Reference

Nash, J. C. (1979) *Compact numerical methods for computers: linear algebra and function minimisation.* New York: Halsted Press.

Example

Sample input

```
178  10      1
SAMPLE RUN OF BIVARIATE TIME DOMAIN ANALYSIS PROGRAM
(25F3.0)
  0  1  9  1  0 -2  1  3 -1 -1  0  2 -2 -2 -1  1  1  0 -5  1  3  2  0  1  1
  1  1  2  0 -3 -2 -2  4  1 -1  1  3  1  3 -4 -2 -4  6  3  0 -1 -4  2 -4  0
  0  0  2  2  0 -2 -1  0  1  5 -3  3 -4 -1 -1 -1 -1 -1 -2 -1 -1  1  1 -2  2
 -1  3  0 -1  0 -1 -3 -1  0  0 -1 -1 -4 -1  5  2  0  2  0  0  0  6 -1  3
 -4 -4  0 -2  4 -1  3  0  0  0  1 -1 -8  0 -1  1 -2  1  1 -1 -1 -1  0  1
  0  7  2 -3 -4  1  1 -1  0 -2  0 -1  0  0  0  2 -1 -1 -1  0 -1 -1  0
  2  0  2  0  4  1  5  0 -5 -1 -1  1  2  0 -1 -1  1 -1 -1  0 -1  0
  2  0  0  0  3 -5  4  0 -1  0 -2 -3 -3  0  1  0 -2  1 -1 -1  0
 -4 -4  6  1 -3  2  3  7 -1 -3  0 -1  3  2  1  0 -1 -4  0  0  2  2  0
  0  2  3  0  1  0  0  0  0  0  0  0  0  0  0  2  1  0  0 -2  2  0
  0  1 -4  0  0 -6  0  0  0  0 -1 -4  0  0 -6 -3 -2  0 -4 -1  0
  0  0  5  1  0  0  0  0 -1  0  1  0  0  0  4  1  0  4  0  0
  0  0  2  0  2 -1 -2  0 -1  1  0  0  0 -1  0 -1 -1  0  0  0
  2  1  1  2 -1 -3  0  0  6 -2  0 -2  0  0  0 -1  0  0  0
  0  1  0  1  0  2 -1 -1  0 -3  0  0  0  0  0  0  0  0
  0  0  0  0  5  0  0  0 -2  0 -5  0  0  0  0  0  0  0
```

Sample output

BIVAR PROGRAM

SAMPLE RUN OF BIVARIATE TIME DOMAIN ANALYSIS PROGRAM

```
178 DATA POINTS
  INITIAL VALUE OF A, B, C, AND D IS    10
  CRITICAL VALUE IS 1.60
  1 FORMAT CARDS
  DATA FORMAT: (25F3.0)
  THE FIRST DATA POINTS OF EACH SERIES ARE:     0.0000    -4.0000
  THE LAST DATA POINTS OF EACH SERIES ARE:      0.0000     0.0000
```

T = 168

A = 10   B = 10   SSE = 674.884   MSE = 4.560   T LN(SSE/T) = 233.617

TERM		PARAMETER VALUE	T STATISTIC
Y1 LAGGED	1 UNITS	-.179	-2.141
Y1 LAGGED	2 UNITS	-.186	-2.128
Y1 LAGGED	3 UNITS	-.324	-3.603
Y1 LAGGED	4 UNITS	-.225	-2.452
Y1 LAGGED	5 UNITS	-.076	-.808
Y1 LAGGED	6 UNITS	-.102	-1.057
Y1 LAGGED	7 UNITS	-.114	-1.217
Y1 LAGGED	8 UNITS	-.057	-.645
Y1 LAGGED	9 UNITS	-.049	-.566
Y1 LAGGED	10 UNITS	-.019	-.228
Y2 LAGGED	1 UNITS	-.010	-.099
Y2 LAGGED	2 UNITS	.235	2.272
Y2 LAGGED	3 UNITS	.131	1.266

TERM	PARAMETER VALUE	T STATISTIC
Y2 LAGGED 4 UNITS	.057	.562
Y2 LAGGED 5 UNITS	-.067	-.653
Y2 LAGGED 6 UNITS	-.068	-.709
Y2 LAGGED 7 UNITS	-.104	-1.111
Y2 LAGGED 8 UNITS	-.152	-1.644
Y2 LAGGED 9 UNITS	.003	.030
Y2 LAGGED 10 UNITS	-.094	-1.052

A = 9   B = 9   SSE = 680.894   MSE = 4.539   T LN(SSE/T) = 235.106

TERM	PARAMETER VALUE	T STATISTIC
Y1 LAGGED 1 UNITS	-.179	-2.149
Y1 LAGGED 2 UNITS	-.173	-2.004
Y1 LAGGED 3 UNITS	-.305	-3.473
Y1 LAGGED 4 UNITS	-.209	-2.315
Y1 LAGGED 5 UNITS	-.060	-.653
Y1 LAGGED 6 UNITS	-.092	-.984
Y1 LAGGED 7 UNITS	-.096	-1.086
Y1 LAGGED 8 UNITS	-.052	-.608
Y1 LAGGED 9 UNITS	-.037	-.437
Y2 LAGGED 1 UNITS	-.005	-.050
Y2 LAGGED 2 UNITS	.236	2.290
Y2 LAGGED 3 UNITS	.124	1.218
Y2 LAGGED 4 UNITS	.064	.633
Y2 LAGGED 5 UNITS	-.082	-.851
Y2 LAGGED 6 UNITS	-.072	-.768
Y2 LAGGED 7 UNITS	-.115	-1.233
Y2 LAGGED 8 UNITS	-.148	-1.605
Y2 LAGGED 9 UNITS	-.004	-.044

A = 8   B = 8   SSE = 681.852   MSE = 4.486   T LN(SSE/T) = 235.343

TERM			PARAMETER VALUE	T STATISTIC
Y1	LAGGED	1 UNITS	-.176	-2.145
Y1	LAGGED	2 UNITS	-.167	-1.985
Y1	LAGGED	3 UNITS	-.300	-3.483
Y1	LAGGED	4 UNITS	-.203	-2.300
Y1	LAGGED	5 UNITS	-.050	-.560
Y1	LAGGED	6 UNITS	-.077	-.885
Y1	LAGGED	7 UNITS	-.088	-1.028
Y1	LAGGED	8 UNITS	-.043	-.519
Y2	LAGGED	1 UNITS	-.006	-.057
Y2	LAGGED	2 UNITS	.230	2.262
Y2	LAGGED	3 UNITS	.122	1.209
Y2	LAGGED	4 UNITS	.048	.509
Y2	LAGGED	5 UNITS	-.090	-.962
Y2	LAGGED	6 UNITS	-.075	-.808
Y2	LAGGED	7 UNITS	-.118	-1.271
Y2	LAGGED	8 UNITS	-.156	-1.740

A = 7   B = 8   SSE = 683.062   MSE = 4.464   T LN(SSE/T) = 235.640

TERM			PARAMETER VALUE	T STATISTIC
Y1	LAGGED	1 UNITS	-.171	-2.108
Y1	LAGGED	2 UNITS	-.162	-1.947
Y1	LAGGED	3 UNITS	-.295	-3.458
Y1	LAGGED	4 UNITS	-.192	-2.246
Y1	LAGGED	5 UNITS	-.034	-.410
Y1	LAGGED	6 UNITS	-.068	-.802
Y1	LAGGED	7 UNITS	-.079	-.942
Y2	LAGGED	1 UNITS	-.012	-.122
Y2	LAGGED	2 UNITS	.227	2.241
Y2	LAGGED	3 UNITS	.104	1.101

```
Y2 LAGGED 4 UNITS              .038              .415
Y2 LAGGED 5 UNITS             -.092             -.988
Y2 LAGGED 6. UNITS            -.078             -.851
Y2 LAGGED 7 UNITS            -.127            -1.407
Y2 LAGGED 8 UNITS            -.166            -1.911

A =  6   B =  8    SSE = 687.019    MSE =  4.461    T LN(SSE/T) = 236.611

   TERM          PARAMETER VALUE        T STATISTIC
Y1 LAGGED 1 UNITS        -.166            -2.047
Y1 LAGGED 2 UNITS        -.156            -1.880
Y1 LAGGED 3 UNITS        -.277            -3.333
Y1 LAGGED 4 UNITS        -.168            -2.060
Y1 LAGGED 5 UNITS        -.021             -.250
Y1 LAGGED 6 UNITS        -.052             -.619
Y2 LAGGED 1 UNITS        -.019             -.188
Y2 LAGGED 2 UNITS         .201             2.066
Y2 LAGGED 3 UNITS         .088              .949
Y2 LAGGED 4 UNITS         .038              .409
Y2 LAGGED 5 UNITS        -.102            -1.096
Y2 LAGGED 6 UNITS        -.095            -1.051
Y2 LAGGED 7 UNITS        -.145            -1.633
Y2 LAGGED 8 UNITS        -.162            -1.862

A =  5   B =  8    SSE = 688.731    MSE =  4.443    T LN(SSE/T) = 237.029

   TERM          PARAMETER VALUE        T STATISTIC
Y1 LAGGED 1 UNITS        -.164            -2.023
Y1 LAGGED 2 UNITS        -.145            -1.794
Y1 LAGGED 3 UNITS        -.263            -3.299
Y1 LAGGED 4 UNITS        -.160            -1.993
Y1 LAGGED 5 UNITS        -.011             -.133
```

TERM	PARAMETER VALUE	T STATISTIC
Y2 LAGGED 1 UNITS	-.036	-.365
Y2 LAGGED 2 UNITS	.192	1.999
Y2 LAGGED 3 UNITS	.089	.953
Y2 LAGGED 4 UNITS	-.032	.354
Y2 LAGGED 5 UNITS	-.113	-1.242
Y2 LAGGED 6 UNITS	-.106	-1.206
Y2 LAGGED 7 UNITS	-.142	-1.605
Y2 LAGGED 8 UNITS	-.164	-1.894

A = 4   B = 8   SSE = 688.809   MSE = 4.415   T LN(SSE/T) = 237.048

TERM	PARAMETER VALUE	T STATISTIC
Y1 LAGGED 1 UNITS	-.162	-2.037
Y1 LAGGED 2 UNITS	-.143	-1.826
Y1 LAGGED 3 UNITS	-.261	-3.331
Y1 LAGGED 4 UNITS	-.159	-2.002
Y2 LAGGED 1 UNITS	-.038	-.392
Y2 LAGGED 2 UNITS	.192	2.008
Y2 LAGGED 3 UNITS	.087	.947
Y2 LAGGED 4 UNITS	.031	.339
Y2 LAGGED 5 UNITS	-.115	-1.301
Y2 LAGGED 6 UNITS	-.105	-1.203
Y2 LAGGED 7 UNITS	-.142	-1.621
Y2 LAGGED 8 UNITS	-.163	-1.897

A = 4   B = 0   SSE = 759.273   MSE = 4.630   T LN(SSE/T) = 253.411

TERM	PARAMETER VALUE	T STATISTIC
Y1 LAGGED 1 UNITS	-.114	-1.473
Y1 LAGGED 2 UNITS	-.084	-1.104
Y1 LAGGED 3 UNITS	-.214	-2.821
Y1 LAGGED 4 UNITS	-.112	-1.452

A = 10   B = 0   SSE = 745.232   MSE = 4.717   T LN(SSE/T) = 250.275

TERM	PARAMETER VALUE	T STATISTIC
Y1 LAGGED 1 UNITS	-.130	-1.638
Y1 LAGGED 2 UNITS	-.108	-1.347
Y1 LAGGED 3 UNITS	-.243	-3.045
Y1 LAGGED 4 UNITS	-.145	-1.780
Y1 LAGGED 5 UNITS	-.044	-.538
Y1 LAGGED 6 UNITS	-.094	-1.145
Y1 LAGGED 7 UNITS	-.085	-1.039
Y1 LAGGED 8 UNITS	-.073	-.965
Y1 LAGGED 9 UNITS	-.050	-.661
Y1 LAGGED 10 UNITS	-.066	-.874

SUMMARY TABLE

T = 168

MODEL	A	B	SSE	T LN(SSE/T)
1	10	10	674.884	233.617
2	4	8	688.809	237.048
3	4	0	759.273	253.411
4	10	0	745.232	250.275
5	0	0	810.000	264.276

1 VS 2	Q = 3.431	8 = DF
2 VS 3	Q = 16.363	8 = DF
3 VS 4	Q = 3.136	6 = DF

C = 10   D = 10   SSE = 412.072   MSE = 2.784   T LN(SSE/T) = 150.735

TERM		PARAMETER VALUE	T STATISTIC
Y2 LAGGED	1 UNITS	-.087	-1.074
Y2 LAGGED	2 UNITS	.008	.104
Y2 LAGGED	3 UNITS	.015	.187
Y2 LAGGED	4 UNITS	-.184	-2.311
Y2 LAGGED	5 UNITS	-.174	-2.156
Y2 LAGGED	6 UNITS	-.106	-1.421
Y2 LAGGED	7 UNITS	.023	.311
Y2 LAGGED	8 UNITS	-.094	-1.303
Y2 LAGGED	9 UNITS	-.110	-1.525
Y2 LAGGED	10 UNITS	-.068	-.976
Y1 LAGGED	1 UNITS	.169	2.587
Y1 LAGGED	2 UNITS	.174	2.551
Y1 LAGGED	3 UNITS	.031	.440
Y1 LAGGED	4 UNITS	.166	2.315
Y1 LAGGED	5 UNITS	.268	3.639
Y1 LAGGED	6 UNITS	.124	1.649
Y1 LAGGED	7 UNITS	.162	2.206
Y1 LAGGED	8 UNITS	.044	.639
Y1 LAGGED	9 UNITS	.078	1.147
Y1 LAGGED	10 UNITS	.112	1.696

C = 9   D = 10   SSE = 414.726   MSE = 2.783   T LN(SSE/T) = 151.814

TERM		PARAMETER VALUE	T STATISTIC
Y2 LAGGED	1 UNITS	-.082	-1.011
Y2 LAGGED	2 UNITS	-.010	.122
Y2 LAGGED	3 UNITS	.016	.193
Y2 LAGGED	4 UNITS	-.177	-2.232
Y2 LAGGED	5 UNITS	-.172	-2.136
Y2 LAGGED	6 UNITS	-.103	-1.381

TERM	PARAMETER VALUE	T STATISTIC
Y2 LAGGED 7 UNITS	.017	.233
Y2 LAGGED 8 UNITS	−.089	−1.230
Y2 LAGGED 9 UNITS	−.109	−1.513
Y1 LAGGED 1 UNITS	.168	2.567
Y1 LAGGED 2 UNITS	.180	2.662
Y1 LAGGED 3 UNITS	.040	.573
Y1 LAGGED 4 UNITS	.174	2.436
Y1 LAGGED 5 UNITS	.275	3.746
Y1 LAGGED 6 UNITS	.124	1.641
Y1 LAGGED 7 UNITS	.165	2.247
Y1 LAGGED 8 UNITS	.041	.604
Y1 LAGGED 9 UNITS	.080	1.183
Y1 LAGGED 10 UNITS	.098	1.524

C = 8   D = 9   SSE = 425.526   MSE = 2.818   T LN(SSE/T) = 156.133

TERM	PARAMETER VALUE	T STATISTIC
Y2 LAGGED 1 UNITS	−.073	−.900
Y2 LAGGED 2 UNITS	.008	.103
Y2 LAGGED 3 UNITS	.042	.524
Y2 LAGGED 4 UNITS	−.169	−2.122
Y2 LAGGED 5 UNITS	−.123	−1.618
Y2 LAGGED 6 UNITS	−.090	−1.222
Y2 LAGGED 7 UNITS	.028	.388
Y2 LAGGED 8 UNITS	−.079	−1.084
Y1 LAGGED 1 UNITS	.173	2.652
Y1 LAGGED 2 UNITS	.183	2.722
Y1 LAGGED 3 UNITS	.034	.490
Y1 LAGGED 4 UNITS	.169	2.401
Y1 LAGGED 5 UNITS	.258	3.538
Y1 LAGGED 6 UNITS	.098	1.334
Y1 LAGGED 7 UNITS	.121	1.755
Y1 LAGGED 8 UNITS	.024	.358

Y1 LAGGED 9 UNITS   .039   .599

C = 7   D = 8   SSE = 429.269   MSE = 2.806   T LN(SSE/T) = 157.604

TERM	PARAMETER VALUE	T STATISTIC
Y2 LAGGED 1 UNITS	-.075	-.918
Y2 LAGGED 2 UNITS	.021	.257
Y2 LAGGED 3 UNITS	.045	.562
Y2 LAGGED 4 UNITS	-.148	-1.990
Y2 LAGGED 5 UNITS	-.120	-1.618
Y2 LAGGED 6 UNITS	-.084	-1.153
Y2 LAGGED 7 UNITS	.031	.423
Y1 LAGGED 1 UNITS	.178	2.778
Y1 LAGGED 2 UNITS	.184	2.780
Y1 LAGGED 3 UNITS	.034	.507
Y1 LAGGED 4 UNITS	.163	2.338
Y1 LAGGED 5 UNITS	.249	3.538
Y1 LAGGED 6 UNITS	.078	1.136
Y1 LAGGED 7 UNITS	.114	1.680
Y1 LAGGED 8 UNITS	.000	.007

C = 6   D = 7   SSE = 429.796   MSE = 2.773   T LN(SSE/T) = 157.810

TERM	PARAMETER VALUE	T STATISTIC
Y2 LAGGED 1 UNITS	-.076	-.956
Y2 LAGGED 2 UNITS	.018	.229
Y2 LAGGED 3 UNITS	.042	.567
Y2 LAGGED 4 UNITS	-.146	-2.016
Y2 LAGGED 5 UNITS	-.119	-1.627
Y2 LAGGED 6 UNITS	-.084	-1.158
Y1 LAGGED 1 UNITS	.176	2.794
Y1 LAGGED 2 UNITS	.181	2.791

			PARAMETER VALUE	T STATISTIC
Y1 LAGGED	3	UNITS	.035	.530
Y1 LAGGED	4	UNITS	.163	2.426
Y1 LAGGED	5	UNITS	.251	3.845
Y1 LAGGED	6	UNITS	.078	1.157
Y1 LAGGED	7	UNITS	.120	1.856

C = 5   D = 7   SSE = 433.515   MSE = 2.779   T LN(SSE/T) = 159.258

TERM			PARAMETER VALUE	T STATISTIC
Y2 LAGGED	1	UNITS	-.070	-.876
Y2 LAGGED	2	UNITS	.033	.423
Y2 LAGGED	3	UNITS	.041	.556
Y2 LAGGED	4	UNITS	-.147	-2.020
Y2 LAGGED	5	UNITS	-.120	-1.636
Y1 LAGGED	1	UNITS	.181	2.882
Y1 LAGGED	2	UNITS	.177	2.731
Y1 LAGGED	3	UNITS	.030	.450
Y1 LAGGED	4	UNITS	.152	2.285
Y1 LAGGED	5	UNITS	.251	3.833
Y1 LAGGED	6	UNITS	.059	.909
Y1 LAGGED	7	UNITS	.105	1.657

C = 5   D = 0   SSE = 501.288   MSE = 3.075   T LN(SSE/T) = 183.660

TERM			PARAMETER VALUE	T STATISTIC
Y2 LAGGED	1	UNITS	.023	.297
Y2 LAGGED	2	UNITS	.082	1.073
Y2 LAGGED	3	UNITS	.097	1.324
Y2 LAGGED	4	UNITS	-.072	-.990
Y2 LAGGED	5	UNITS	-.061	-.832

C = 10   D = 0   SSE = 479.088   MSE = 3.032   T LN(SSE/T) = 176.050

TERM	PARAMETER VALUE	T STATISTIC
Y2 LAGGED 1 UNITS	.003	.041
Y2 LAGGED 2 UNITS	.059	.764
Y2 LAGGED 3 UNITS	.089	1.200
Y2 LAGGED 4 UNITS	-.084	-1.147
Y2 LAGGED 5 UNITS	-.055	-.759
Y2 LAGGED 6 UNITS	-.063	-.870
Y2 LAGGED 7 UNITS	.046	.639
Y2 LAGGED 8 UNITS	-.118	-1.683
Y2 LAGGED 9 UNITS	-.102	-1.458
Y2 LAGGED 10 UNITS	-.076	-1.090

SUMMARY TABLE

T = 168

MODEL	C	D	SSE	T LN(SSE/T)
1	10	10	412.072	150.735
2	5	7	433.515	159.258
3	5	0	501.288	183.660
4	10	0	479.088	176.050
5	0	0	515.000	188.194

1 VS 2   Q = 8.522   8 = DF
2 VS 3   Q = 24.403   7 = DF
3 VS 4   Q = 7.610   5 = DF

END OF BIVAR PROGRAM

TSREG

Program description

The program TSREG performs the two-stage least-squares regression procedure as described in Chapter 27 of Gottman's *Time-Series Analysis*.

Program steps

The program user supplies the dependent variable time-series $(\mathbf{Y})$ and k independent variable time-series $(\mathbf{X})$. Each time-series must have the same number of data points (N).

The dependent variable time-series is described by the matrices:

$$\begin{bmatrix} Y_1 \\ Y_2 \\ \vdots \\ Y_N \end{bmatrix} = \begin{bmatrix} 1 & X_{11} & X_{21} & \cdots & X_{k1} \\ 1 & X_{12} & X_{22} & \cdots & X_{k2} \\ \vdots & \vdots & \vdots & & \vdots \\ 1 & X_{1N} & X_{2N} & \cdots & X_{kN} \end{bmatrix} \begin{bmatrix} \beta_0 \\ \beta_1 \\ \vdots \\ \beta_k \end{bmatrix} + \begin{bmatrix} e_1 \\ e_2 \\ \vdots \\ e_N \end{bmatrix}.$$

In stage one, the least-squares parameter estimates $(\mathbf{B})$ are computed by the matrix equation:

$$\hat{\mathbf{B}} = (\mathbf{X}^T\mathbf{X})^{-1}\mathbf{X}^T\mathbf{Y} \ .$$

There are $k+1$ parameter estimates in $(\mathbf{B})$. For all matrix inversion in this program, the Bauer-Reinsch method is used as described in Nash (1979).

The estimated residual series of stage one is:

$$\hat{\mathbf{E}} = (\mathbf{Y}-\mathbf{X}\hat{\mathbf{B}}) \ ,$$

in which there are N elements. The mean of this series is computed as:

$$\bar{E} = \frac{1}{N}\sum_{i=1}^{N} e_i \ ,$$

and the variance is:

$$C_0 = \frac{1}{N}\sum_{i=1}^{N} (e_i-\bar{E})^2 \ .$$

The autocovariances of the residual series is computed:

$$C_i = \frac{1}{N} \sum_{m=1}^{N-i} (e_m - \bar{E})(e_{m+i} - \bar{E})$$

for all i from 1 to $(N-1)$. The autocorrelations are calculated with the equation:

$$R_i = C_i/C_0 .$$

An option is provided in the program for the autoregressive model fit of the order $N/10$ on the residual time-series $(\hat{E})$. If this is requested by the user, all of the calculations are performed as described for the ARFIT program, except $(\hat{E})$ is the data input instead of (X).

For stage two, the striped matrix (OMEGA) is made up of the auto-correlations (R) in the following way:

$$\Omega = \begin{bmatrix} 1 & r_1 & r_2 & r_3 & \cdots & r_{N-1} \\ r_1 & 1 & r_1 & r_2 & \cdots & r_{N-2} \\ r_2 & r_1 & 1 & r_1 & \cdots & r_{N-3} \\ \vdots & \vdots & \vdots & \vdots & & \vdots \\ r_{N-1} & r_{N-2} & r_{N-3} & r_{N-4} & \cdots & 1 \end{bmatrix} .$$

The stage two least-squares parameter estimates $(\hat{B}_*)$ are computed by:

$$\hat{B}_* = (X^T \Omega^{-1} X)^{-1} X^T \Omega^{-1} Y .$$

There are $(k+1)$ parameter estimates.

The residual time-series for stage two is:

$$\hat{E}_* = Y - X\hat{B}_* .$$

There are N elements in the residual time-series for stage two.

The second stage residual variance is:

$$\sigma_*^2 = \hat{E}_*^T \Omega^{-1} \hat{E}_* / (N-k-1) ,$$

and the standard error is the square root of the variance.

The variance for the least-squares parameter estimates $(\hat{B}_*)$ is in the diagonal of the variance-covariance matrix:

$$\sigma_*^2 (X^T \Omega^{-1} X)^{-1} .$$

The i th diagonal element is referred to as CII in the program.

The confidence interval of $(\hat{B}_*)$ is determined by:

$$\hat{\beta}_i \pm (t_{\alpha/2}) \sqrt{\text{CII}_i} \, ,$$

in which $t_{\alpha/2}$ is the $\alpha/2$ level of Student's t-statistic with $(N-k-1)$ degrees of freedom. If degrees of freedom are less than or equal to 30, this calculation is done by the program.

The t-ratio is also calculated for each beta in $(\hat{\mathbf{B}}_.)$:

$$t_i = \frac{\beta_i}{\sqrt{\text{CII}_i}}$$

with $(N-k-1)$ degrees of freedom.

The variance of $(\mathbf{Y})$ minus the second stage residual variance over the variance of $(\mathbf{Y})$ yields the percent of variance accounted for.

Program input

1. First input card.
   - Cols. 1-5   Number of data points (N) in each time-series. Maximum is 150.
   - Cols. 9-10   Number of independent variable series (k). Maximum is 10.
   - Col. 15   Set to 1 if autoregressive model fit is to be performed on the residual series of stage one.
   - Col. 18   Set to 1 if the input is the first data point for all time-series followed by the second data point for all time-series (e.g., $Y_1, X_{11}, X_{12}, \cdots, X_{1k}, Y_2, X_{21}, X_{22}, \cdots, X_{2k}$, etc.). Otherwise all data for the dependent variable series (Y) will be read, followed by all data for the first independent variable, then the second independent variable and so on.
   - Col. 20   Number of format cards to follow. Maximum is 5.
   - Col. 25   Set to 1 if the second stage residual series is to go to the punch.
2. Second input card.
   - Cols. 1-80   The title which will label the output.
3. Next input card or cards.
   - Cols. 1-80   Data format is specified on this card or cards. Standard FORTRAN F-format is used, enclosed in parentheses.
4. Time-series data cards.
   - Cols. 1-80   Data cards for the time-series follow the last format card. Data are punched in the format specified on the format card or cards and as specified by column 18 of the first input card.

88

The dependent variable must precede the independent variables whether the data are read by rows or by columns.

Program output

When the program is run, the following information is output:
1. Program name: TSREG PROGRAM.
2. The title specified for printout identification.
3. Number of data points (N) specified.
4. Specified number of independent variable time-series (k).
5. Number of format cards specified.
6. Data format is repeated back to the user.
7. The first and last data points of the dependent variable time-series are printed and the first and last data points of all independent variable time-series are printed for the user to check the format.
8. The autocorrelations (R) for the first stage residuals are printed.
9. If the autoregressive model fit is requested, a matrix of the residual variance, autoregressive model coefficients, and partial autocorrelation coefficients for each order of the model up to $N/10$ is printed.
10. If the autoregressive model fit is requested, a table of the coefficients of the final model with the standard deviation for each coefficient and Student's t-ratio with degrees of freedom is printed.
11. Second stage residual variance and standard error.
12. A table showing the second stage least-squares parameter estimates ($\hat{\mathbf{B}}$), variance for the estimates, degrees of freedom, and the 95% and 99% confidence intervals. The confidence intervals are not printed if degrees of freedom are greater than 30. In the first row of this table is β_0, which is the constant or intercept in the linear representation. The other estimates are the coefficients of the corresponding independent variables.
13. The t-ratio is given for each β for significance testing.
14. The percent of variance accounted for.
15. The second stage residual time-series.
16. The end of program message.

Special notes

1. If the number of data points is required to be greater than 125 or the number of independent variables is required to be greater than 10, array dimensions in the program will have to be increased and the statements near the beginning of the pro-

gram which check for these maxima will have to be modified. If N is the number of data points and k is the number of independent variables, the arrays are dimensioned as follows: $Y(N)$, $X(N,k+1)$, $B(k+1)$, $XT(k+1,N)$, $XINV(N(N+1)/2)$, $TEMP(k+1,N)$, $E(N)$, $C(N-1)$, $R(N-1)$, $A(^{N}/_{10}(^{N}/_{10}+1)/2)$, $RES(N)$, $YINV(^{N}/_{10}(^{N}/_{10}+1)/2)$, $XTOY(k+1)$, $OMEGA(N(N+1)/2)$.

2. See Special Notes 2, 3 and 4 for program DETRND.

References

Goldberger, A. S. (1964) *Econometric theory*. New York: Wiley, pp. 231-235.

Johnston, J. (1972) *Econometric methods* (2nd ed.). New York: McGraw-Hill, p. 259.

Nash, J. C. (1979) *Compact numerical methods for computers: linear algebra and function minimisation*. New York: Halsted Press.

Example

Sample input

```
   30       3      1  1  1      1
SAMPLE RUN OF THE TWO-STAGE LEAST-SQUARES PROGRAM
(4F10.2)
        861.90    3263.00     266.40     198.50
        891.20    3002.60     284.80     216.10
        926.90    3534.00     266.30     234.80
        902.10    3193.00     260.90     231.20
        855.60    3081.00     273.50     233.20
        784.60    3159.80     253.30     226.50
        940.40    3764.80     260.00     296.20
        920.80    4292.30     250.90     227.50
        864.70    3267.00     265.10     232.40
        867.90    3554.50     272.10     237.00
        852.70    3205.80     268.90     236.20
        752.40    2677.00     265.40     221.90
        869.80    2572.00     255.90     228.10
        872.10    3140.00     249.60     234.50
        851.60    3208.00     249.80     233.50
        852.40    3280.20     259.20     235.00
        867.50    2583.40     260.80     241.20
        871.80    3405.00     255.80     237.20
        816.40    3129.40     277.50     245.10
        768.60    2661.40     253.10     232.40
        863.40    3259.80     261.60     236.60
        849.20    3616.00     255.20     237.10
        826.10    3254.00     263.30     237.60
        812.30    2798.20     260.50     238.20
        844.10    2578.00     260.60     241.30
        821.60    3017.40     259.20     232.00
        864.00    3795.50     259.40     234.30
        837.20    3150.30     275.30     242.40
        848.60    3651.00     266.10     235.30
        879.00    4090.30     260.40     231.10
```

Sample output

TSREG PROGRAM

SAMPLE RUN OF THE TWO-STAGE LEAST-SQUARES PROGRAM

    30 DATA POINTS IN EACH SERIES
     3 INDEPENDENT VARIABLE SERIES
     1 FORMAT CARDS
     DATA FORMAT: (4F10.2)

THE RESIDUAL SERIES WILL GO TO PUNCH
    PUNCH FORMAT IS (I5,F10.4)

DATA WILL BE READ ONE POINT FOR EACH TIME-SERIES
    FOLLOWED BY NEXT POINT FOR EACH SERIES AND SO ON.

FIRST DATA POINT OF DEPENDENT VARIABLE SERIES: 861.9000
LAST DATA POINT OF DEPENDENT VARIABLE SERIES: 879.0000
FIRST AND LAST DATA POINTS OF ALL OTHER SERIES ARE: 3263.0000   4090.3000
                                                       266.4000   260.4000
                                                       198.5000   231.1000

AUTOCORRELATIONS FOR THE FIRST STAGE RESIDUALS

ORDER	R	ORDER	R	ORDER	R	ORDER	R
1	.255	2	-.054	3	-.004	4	.172
5	-.082	6	.082	7	-.014	8	.032
9	-.034	10	.097	11	.007	12	.041
13	.036	14	.136	15	-.091	16	-.108
17	-.049	18	-.061	19	-.174	20	-.109
21	-.077	22	-.043	23	-.009	24	-.054
25	-.127	26	-.122	27	-.088	28	-.046
29	-.012						

AUTOREGRESSIVE MODEL FIT TO FIRST STAGE RESIDUALS

THE AUTOREGRESSIVE COEFFICIENTS,
PARTIAL AUTOCORRELATION COEFFICIENTS AND RESIDUAL VARIANCE

ORDER	RES. VAR.	A(1)	A(2)	A(3)	A(4)	A(
1	1037.486	.255				
2	1020.615	.287	-.128			
3	1018.281	.294	-.141	.048		
4	989.711	.286	-.118	-.001	.168	

STANDARD DEVIATION OF AR PARAMETER ESTIMATES
AND STUDENT'S T-RATIO WITH 26 DEGREES OF FREEDOM

	1	2	3	4
COEFFICIENTS	.286	-.118	-.001	.168
SD	.168	.175	.175	.168
T-RATIO	1.700	-.673	-.008	.998

SECOND STAGE RESIDUAL VARIANCE IS 1128.4762
AND STANDARD ERROR IS 33.5928

BETA ESTIMATES	VARIANCE	DF	95% CONFIDENCE INT.		99% CONFIDENCE INT.	
0	468.2668	27244.7066	26	75.6228 --	860.9109	-37.9548 --	974.4884
1	.0519	.0002	26	.0212 --	.0826	.0123 --	.0914
2	.2722	.3540	26	-1.1432 --	1.6876	-1.5526 --	2.0970
3	.6256	.1165	26	-.1864 --	1.4376	-.4213 --	1.6725

```
BETA    T-RATIO -- 26 DEGREES OF FREEDOM
0    2.8370
1    4.0210
2     .4575
3    1.8327

PERCENT OF VARIANCE ACCOUNTED FOR IS    33.68

    RESIDUAL TIME SERIES
 1   27.6600    2   54.4497    3   55.9188    4   52.5311
 5    7.1604    6  -58.2375    7   20.7480    8   19.2391
 9    9.3982   10   -7.0997   11   -2.8386   12  -65.8070
13   55.7473   14   26.2919   15    2.8354   16   -3.6072
17   43.3266   18    8.8676   19  -43.0840   20  -52.0185
21    6.7968   22  -24.4526   23  -31.2906   24  -21.0581
25   20.1986   26  -18.8970   27  -18.3561   28  -21.0803
29  -28.7092   30  -16.9198

END OF TSREG PROGRAM
```

CRSPEC

Program description

The program CRSPEC performs a cross-spectral analysis on two time-series. The procedure is discussed in Chapter 23 of Gottman's *Time-Series Analysis*.

Program steps

The program user supplies the data for the two time-series (X) and (Y). The number of points (N) in each series must be equal. The maximum number of lags $(MLAG)$ to be used in the computations is specified by the user and is usually around $N/6$.

If the number of data points (N) is even, the program drops the last data point from each series so N becomes odd.

The mean of each series is calculated. The mean of (X) is subtracted from each data point x, and the mean of (Y) is subtracted from each y. The autocovariances are then calculated:

$$CX_0 = \frac{1}{N}\sum_{t=1}^{N}(x_t-\bar{X})^2 ,$$

$$CY_0 = \frac{1}{N}\sum_{t=1}^{N}(y_t-\bar{Y})^2 ,$$

$$CX_k = \frac{1}{N}\sum_{t=1}^{N-k}(x_t-\bar{X})(x_{t+k}-\bar{X}) ,$$

and

$$CY_k = \frac{1}{N}\sum_{t=1}^{N-k}(y_t-\bar{Y})(y_{t+k}-\bar{Y}) .$$

The covariances are calculated for each lag k up to $MLAG$.

The cross-covariances are calculated for all lags k from $-MLAG$ to $MLAG$:

$$CXY_k = \frac{1}{N}\sum_{t=1}^{N-k}(x_t-\bar{X})(y_{t+k}-\bar{Y}) ,$$

$$CXY_{-k} = \frac{1}{N}\sum_{t=1}^{N-|k|}(x_{t+|k|}-\bar{X})(y_t-\bar{Y}) ,$$

and

93

94

$$CXY_0 = \frac{1}{N}\sum_{t=1}^{N}(x_t - \bar{X})(y_t - \bar{Y}) .$$

The user of the program has the option of letting the program shift all of the cross-covariances if CXY_0 isn't the maximum cross-covariance. If this option is specified, the cross-covariance values are shifted either direction in the vector so that the maximum covariance is at CXY_0 and $MLAG$ is reduced by the number of positions of the shift. This number of positions of the shift is called the shift parameter.

The Tukey-Hanning window is employed in the further calculations. Weights are calculated for each lag k from 1 to $MLAG$:

$$W_k = \tfrac{1}{2}(1+\cos\pi (k/MLAG)) .$$

Each spectrum calculated in the program is calculated for the frequencies in the overtone series:

$$F_j = \frac{j}{N} ,$$

in which $j = 1, 2, \cdots , (N-1)/2$. Then the calculations are also done for $F = 0$ and $F = .5$.

The auto-spectral analysis is performed on each series as in the SPEC program:

$$PX_{(F_j)} = \frac{1}{2\pi}\left\{CX_0 + 2\sum_{k=1}^{MLAG} W_k CX_k \cos(k2\pi F_j)\right\}$$

and

$$PY_{(F_j)} = \frac{1}{2\pi}\left\{CY_0 + 2\sum_{k=1}^{MLAG} W_k CY_k \cos(k2\pi F_j)\right\} .$$

The estimated degrees of freedom are:

$$EDF = \frac{8}{3}\frac{N}{MLAG}$$

except when $F = 0$ or $F = .5$, $EDF = EDF/2$.

In the cross-covariance calculation, the cospectrum and quadrature spectrum are calculated. In the program, the cospectrum array is called COSPEC and the quadrature spectrum array is called QUSPEC.

$$COSPEC_{(F_j)} = \frac{1}{2\pi}\left\{\sum_{k=-MLAG}^{MLAG} W_k CXY_k \cos(2\pi F_j k)\right\}$$

and

$$QUSPEC_{(F_j)} = \frac{1}{2\pi}\left\{\sum_{k=-MLAG}^{MLAG} W_k CXY_k \sin(2\pi F_j k)\right\}.$$

When $k = 0$, $W_0 = 1$.

Once the above calculations are completed, the coherence spectrum is computed:

$$RHOSQ_{(F_j)} = \frac{COSPEC^2_{(F_j)} + QUSPEC^2_{(F_j)}}{PX_{(F_j)}PY_{(F_j)}}.$$

The phase spectrum is computed:

$$PHI_{(F_j)} = \arctan\frac{-QUSPEC_{(F_j)}}{COSPEC_{(F_j)}}.$$

The auto-spectra are printed and plotted. The coherence spectrum is printed and plotted with a test for coherence $= 0$. For this test $RHOSQ_{(F_j)}$ is not zero if:

$$RHOSQ_{(F_j)} > \frac{FTAB_{[2,2(NN-1)]}}{(NN-1)(1+FTAB_{[2,2(NN-1)]})},$$

in which $NN = EDF/2$ and $FTAB$ refers to an array in the program of the F-distribution. If $2(NN-1) > 30$, the test is not printed and the user will have to perform the test.

The phase spectrum is printed and plotted with a 95% confidence interval:

$$PHI_{(F_j)} \pm \arcsin\left\{TTAB_{EDF(1-\alpha)}\sqrt{\frac{1}{EDF}\frac{1-RHOSQ_{(F_j)}}{RHOSQ_{(F_j)}}}\right\},$$

where $TTAB$ refers to an array in the program of Student's t-distribution ($\alpha = .05$). If $EDF > 30$ or if the arcsin argument is greater than one, the confidence interval will not be calculated by the program.

Program input

1. First input card.

   Cols. 2-5   Number of data points (N) in each time-series. Maximum is 600.

   Cols. 8-10   The maximum number of lags ($MLAG$) to be

	used in the calculations. Maximum is 300.
Col. 15	Set to 1 if lag shift correction is to be made by program.
Col. 18	Set to 1 if data points for the two series alternate on input (e.g., x_1, y_1, x_2, y_2, etc.). Otherwise all data for the first series (X) will be read followed by all data for the second series (Y) with the second series beginning on a new card if there is more than one data point on a card.
Col. 20	Number of data format cards. Maximum is 5.

2. Second input card.

| Cols. 1-80 | The title to be printed for identification of output. |

3. Next input card or cards.

| Cols. 1-80 | Time-series data format is specified on this card or cards. Standard FORTRAN F-format is used, enclosed in parentheses. |

4. Time-series data cards.

| Cols. 1-80 | Data cards for the time-series follow the last format card. Data are punched in the format specified on the data format card or cards and as specified by column 18 of the first input card. |

Program output

When the program is run, the following information is output:

1. Program name: CRSPEC PROGRAM.
2. The title for the printout identification.
3. The specified number of data points (N) in each time-series.
4. The specified maximum number of lags to be used $(MLAG)$.
5. The specified number of format cards.
6. The specified data format is printed back for the user.
7. First and last data points of each time-series are printed so that the user can check the input format.
8. Table of autocovariances (CX) for the first series for lags 0 to $MLAG$.
9. Table of autocovariances (CY) for the second series for lags 0 to $MLAG$.
10. Table of cross-covariances (CXY) for lags $-MLAG$ to $MLAG$ after shift, if shift was performed.
11. The shift parameter is printed.
12. Auto-spectral density estimates (PX) for first time-series for all

frequencies.
13. Plot of the density estimates (*PX*).
14. Auto-spectral density estimates (*PY*) for second time-series for all frequencies.
15. The coherence spectrum (*RHOSQ*) and test for coherence = 0.
16. The plot of the coherence spectrum and test for coherence = 0.
17. The phase spectrum (*PHI*), degrees of freedom, and the confidence interval at the .05 level.
18. The plot of the phase spectrum and the confidence interval. The confidence interval is indicated by asterisks except where the confidence interval value falls outside of the range of the plot.
19. End of program message.

Special notes

1. If more than 600 data points are required, the dimensions of arrays X and Y must be changed to the required number of data points; dimensions of arrays CXYA, CXYB, and W must be at least as big as maximum number of lags to be used (*MLAG*); dimensions of CX and CY should be *MLAG*+1; and dimensions of PX, PY, COSPEC, QUSPEC, RHOSQ and PHI should be at least *MLAG*−1. Statements near the start of the program which check for these maxima should be modified.
2. See Special Notes 3 and 4 for the program DETRND.

Example

Sample input

```
    60    10          1 1
SAMPLE RUN OF CRSPEC PROGRAM
(2F5.2)
  1.61 1.00
  1.00 1.38
  1.38 2.05
  2.05 2.21
  2.21 2.00
  2.00 1.35
  1.35 2.17
  2.17 2.25
  2.25 2.71
  2.71 3.39
  3.39 2.50
  2.50 3.56
  3.56 4.00
  4.00 3.76
  3.76 3.29
  3.29 3.28
```

```
3.28  2.31
2.31  2.50
2.50  2.33
2.33  2.27
2.27  2.35
2.35  3.33
3.33  1.33
1.33  2.19
2.19  1.44
1.44  1.24
1.24  1.44
1.44  4.40
4.40  4.50
4.50  3.56
3.56  3.00
3.00  2.29
2.29  1.83
1.83  2.24
2.24  1.39
1.39  2.82
2.82  3.29
3.29  3.33
3.33  3.33
3.33  2.71
2.71  2.29
2.29  1.38
1.38  2.83
2.83  1.00
1.00  1.33
1  33  3.28
3.28  3.12
3.12  3.59
3.59  3.24
3.24  2.56
2.56  1.44
1.44  1.17
1.17  2.29
2.29  4.00
4.00  4.47
4.47  4.60
4.60  3.37
3.37  2.42
2.42  2.20
2.20  1.61
```

Sample output

CRSPEC PROGRAM

SAMPLE RUN OF CRSPEC PROGRAM

60 DATA POINTS IN EACH SERIES
NUMBER OF LAGS TO BE USED IS 10
1 FORMAT CARDS
DATA FORMAT: (2F5.2)

DATA WILL BE READ ALTERNATING ONE POINT FOR EACH TIME-SERIES

THE FIRST DATA POINTS OF EACH SERIES ARE: 1.610   1.000
THE LAST DATA POINTS OF EACH SERIES ARE: 2.200   1.610

TABLE OF AUTOCOVARIANCES FOR X

LAG	C	LAG	C	LAG	C	LAG	C	LAG	C
0	.916	1	.516	2	.167	3	-.128	4	-.290
5	-.260	6	-.055	7	.063	8	.125	9	.108
10	-.025								

TABLE OF AUTOCOVARIANCES FOR Y

LAG	C	LAG	C	LAG	C	LAG	C	LAG	C
0	.902	1	.491	2	.141	3	-.151	4	-.308
5	-.279	6	-.074	7	.065	8	.126	9	.110
10	-.017								

TABLE OF CROSS-COVARIANCES

LAG	CXY	LAG	CXY	LAG	CXY	LAG	CXY	LAG	CXY
-10	.109	-9	.118	-8	.055	-7	-.077	-6	-.271
-5	-.296	-4	-.138	-3	.146	-2	.490	-1	.899
0	.517	1	.162	2	-.141	3	-.301	4	-.269
5	-.052	6	.073	7	.133	8	.109	9	-.029
10	-.228								

THE SHIFT PARAMETER IS   0

SPECTRAL DENSITY ESTIMATES FOR X

FREQ	DENSITY	FREQ	DENSITY	FREQ	DENSITY	FREQ	DENSITY	FREQ	DENSITY
0.000	.223	.017	.232	.034	.258	.051	.297	.068	.341
.085	.378	.102	.397	.119	.389	.136	.353	.153	.295
.169	.225	.186	.159	.203	.105	.220	.070	.237	.052
.254	.047	.271	.048	.288	.051	.305	.052	.322	.052
.339	.050	.356	.046	.373	.041	.390	.037	.407	.034
.424	.033	.441	.033	.458	.036	.475	.038	.492	.039
.500	.040								

PLOT OF DENSITY ESTIMATES
DENSITY VALUES APPEAR ACROSS TOP AND FREQUENCIES DOWN
THE LEFT SIDE OF PLOT.

```
       .0327   .0847   .1367   .1887   .2407   .2928   .3448     .396
      +.......+.......+.......+.......+.......+.......+.......+
0.000 +.......+.......+.......+.......+.......+.......+.......+
 .017                                 X
 .034                            X
 .051                                     X
 .068                                              X
 .085                                          X
 .102                                              X
 .119                                                      X
                                                             X
```

```
.136
.153
.169
.186                                                    X
.203
.220
.237 X
.254 X                                            X
.271 X
.288 X
.305 X
.322 X                              X
.339 X
.356 X
.373 X
.390X                  X
.407X
.424X
.441X
.458X
.475X
.492 X        X
.500 X
```

SPECTRAL DENSITY ESTIMATES FOR Y

FREQ	DENSITY	FREQ	DENSITY	FREQ	DENSITY	FREQ	DENSITY	FREQ	DENSITY
0.000	.191	.017	.202	.034	.232	.051	.278	.068	.328
.085	.372	.102	.396	.119	.391	.136	.356	.153	.297
.169	.227	.186	.160	.203	.106	.220	.072	.237	.055
.254	.050	.271	.052	.288	.054	.305	.055	.322	.054
.339	.051	.356	.047	.373	.042	.390	.039	.407	.036
.424	.035	.441	.036	.458	.038	.475	.039	.492	.036
.500	.041								

PLOT OF DENSITY ESTIMATES
DENSITY VALUES APPEAR ACROSS TOP AND FREQUENCIES DOWN
THE LEFT SIDE OF PLOT.

.475X
.492 X
.500 X

COHERENCE SPECTRUM

FREQ	COHERENCE	TEST FOR COHERENCE = 0
0.000	.9553	.2790
.017	.9491	.1127
.034	.9383	.1127
.051	.9346	.1127
.068	.9397	.1127
.085	.9492	.1127
.102	.9589	.1127
.119	.9667	.1127
.136	.9721	.1127
.153	.9746	.1127
.169	.9728	.1127
.186	.9638	.1127
.203	.9435	.1127
.220	.9140	.1127
.237	.8937	.1127
.254	.8881	.1127
.271	.8812	.1127
.288	.8784	.1127
.305	.8889	.1127
.322	.9087	.1127
.339	.9235	.1127
.356	.9184	.1127
.373	.8891	.1127
.390	.8496	.1127
.407	.8288	.1127
.424	.8448	.1127
.441	.8781	.1127
.458	.8964	.1127

PLOT OF COHERENCE SPECTRUM
COHERENCE VALUES APPEAR ACROSS TOP AND FREQUENCIES DOWN
THE LEFT SIDE OF PLOT.
TEST FOR COHERENCE = 0 IS SHOWN BY PLUSES.

```
.475   .8940   .1127
.492   .8861   .1127
.500   .8848   .2790
```

```
     0.0000    .1429    .2857    .4286    .5714    .7143    .8571    1.000
0.000
 .017
 .034
 .051
 .068
 .085
 .102
 .119
 .136
 .153
 .169
 .186
 .203
 .220
 .237
 .254
 .271
 .288
 .305
 .322
 .339
 .356
 .373
```

PHASE SPECTRUM

FREQ	PHASE	EDF	CONFIDENCE INTERVAL AT .05 LEVEL	
0.000	0.0000	8	-.1772 ---	.1772
.017	.1373	16	.0143 ---	.2604
.034	.2711	16	.1348 ---	.4074
.051	.3940	16	.2533 ---	.5346
.068	.4992	16	.3646 ---	.6339
.085	.5852	16	.4623 ---	.7082
.102	.6545	16	.5446 ---	.7645
.119	.7115	16	.6131 ---	.8100
.136	.7610	16	.6711 ---	.8509
.153	.8079	16	.7222 ---	.8936
.169	.8587	16	.7700 ---	.9474
.186	.9234	16	.8205 ---	1.0262
.203	1.0191	16	.8891 ---	1.1492
.220	1.1676	16	1.0042 ---	1.3309
.237	1.3619	16	1.1781 ---	1.5457
.254	1.5422	16	1.3529 ---	1.7315
.271	1.6769	16	1.4811 ---	1.8727
.288	1.7873	16	1.5889 ---	1.9858
.305	1.8900	16	1.7015 ---	2.0785
.322	1.9847	16	1.8159 ---	2.1536
.339	2.0686	16	1.9154 ---	2.2217
.356	2.1469	16	1.9882 ---	2.3055
.373	2.2355	16	2.0472 ---	2.4237
.390				
.407				
.424				
.441				
.458				
.475				
.492				
.500				

.390	2.3546	16	2.1297	--	2.5795
.407	2.5108	16	2.2676	--	2.7541
.424	2.6793	16	2.4501	--	2.9084
.441	2.8238	16	2.6250	--	3.0225
.458	2.9339	16	2.7528	--	3.1151
.475	3.0222	16	2.8387	--	3.2057
.492	3.1023	16	2.9112	--	3.2934
.500	-3.1416	8	-3.4402	--	-2.8430

PLOT OF PHASE SPECTRUM

PHASE VALUES APPEAR ACROSS TOP AND FREQUENCIES DOWN
THE LEFT SIDE OF PLOT.
THE PHASE IS INDICATED BY X AND CONFIDENCE INTERVAL
AT .05 LEVEL IS INDICATED BY ASTERISKS.

```
        -3.1416   -2.2440   -1.3464    -.4488     .4488     1.3464    2.2440    3.141
          +.........+.........+.........+.........+.........+.........+.........+
0.000     +
 .017                                              . x
 .034                                               . x
 .051                                                . x .
 .068                                                 . x .
 .085                                                  . x .
 .102                                                   . x .
 .119                                                    . x .
 .136                                                     . x .
 .153                                                      . x .
 .169                                                       . x .
 .186                                                        . x .
 .203                                                         . x .
 .220                                                          . x .
 .237                                                           . x .
 .254                                                            . x .
 .271                                                             . x .
```

```
.288
.305
.322
.339
.356
.373
.390
.407
.424
.441
.458
.475
.492
.500X    .
```

END OF CRSPEC PROGRAM

The programs are available on IBM card source decks, as individual programs or as packages, from the authors at
P. O. Box 3092
Champaign, Illinois 61820
U. S. A.